Magnetic Skyrmions

Edited by Dipti Ranjan Sahu

Published in London, United Kingdom

IntechOpen

Supporting open minds since 2005

Magnetic Skyrmions
http://dx.doi.org/10.5772/intechopen.92951
Edited by Dipti Ranjan Sahu

Contributors
Devinder Singh, Kuldeep Chand Verma, Rajesh Kumar Rajagopal, Fanny Béron, Jeovani Brandão, Marcos Vincius Puydinger Dos Santos, Lalla Btissam Drissi, El Hassan Saidi, Mosto Bousmina, Omar Fassi-Fehri, Abdelhamid El Kaaouachi, Brahim Ait Hammou, Abdellatif El Oujdi, Adil Echchelh, Said Dlimi, Che-Ti Liang, Jamal Hemine, mosbah jemmali, Lotfi Bessais

First published in London, United Kingdom, 2021 by IntechOpen
IntechOpen is the global imprint of INTECHOPEN LIMITED, registered in England and Wales, registration number: 11086078, 5 Princes Gate Court, London, SW7 2QJ, United Kingdom
Printed in Croatia

British Library Cataloguing-in-Publication Data
A catalogue record for this book is available from the British Library

Additional hard and PDF copies can be obtained from orders@intechopen.com

Magnetic Skyrmions
Edited by Dipti Ranjan Sahu
p. cm.
Print ISBN 978-1-83969-252-9
Online ISBN 978-1-83969-253-6
eBook (PDF) ISBN 978-1-83969-254-3

For EU product safety concerns:
IN TECH d.o.o., Prolaz Marije Krucifikse Kozulić 3, 51000 Rijeka, Croatia, info@intechopen.com or visit our website at intechopen.com.

Meet the editor

Dr. Dipti Ranjan Sahu is an Associate Professor of Physics in the Department of Natural and Applied Sciences at the Namibia University of Science and Technology (NUST). He received his doctoral degree in Physics from the Institute of Materials Science, Utkal University, India. After earning his Ph.D., he worked as a postdoctoral researcher and visiting scientist and served as a faculty member in several institutions such as the National Taiwan University, National Cheng Kung University (Taiwan), and the University of Witwatersrand (South Africa). His research focus is on multifunctional materials including nanomaterials, ceramics, composites, spintronics, magnetic materials, and the application of functional materials in devices. He has published more than 100 peer-reviewed articles, 2 books, 2 book chapters, and more than 100 research articles in conference proceedings and meetings.

Contents

Preface

Skyrmions are generally known as nanoscale magnetic vortices, a type of quasiparticles that are driven by ultra-low electrical current. This skyrmion is a point-like region of reversed magnetization in a uniform magnet. The magnetic skyrmions found in magnetic materials exhibit spiral magnetism. Magnetic skyrmions are anticipated to allow for the existence of discrete magnetic states, which are significantly more energetically stable than their single-domain counterparts. Magnetic skyrmions have important applications in microwave communication, logic computing technology, and memory and logic devices in terms of storing information. Magnetic skyrmions within a film or nanotrack can be manipulated using spin currents or spin waves that can have more applications in data storage and spintronic devices and show outstanding magnetic and transport properties. Magnetic skyrmionics is an advanced and active research field that reveals fundamental physics, encourages the development of next-generation high-density efficient information devices, creates and manipulates nanometer-size skyrmions in devices, and develops compatible materials at room temperature by all-electrical means. There are still several challenges such as the interaction of skyrmions with other magnetic textures, the particle-wave duality of skyrmions, skyrmion lattice phase transitions, and the use of skyrmion lattices as magnonic crystals that are needed to address for designing fully functional and competitive skyrmion devices. This book addresses scientific advances and reveals an interesting fact that skyrmions can create and access magnetic skyrmions under ambient room-temperature conditions for next-generation devices.

This book contains six chapters with some of the most exciting theoretical concepts and reviews the development of magnetic skyrmions and their application in magnetic materials. The first three chapters discuss magnetic skyrmions in solids, thin films, and multilayers. The fourth chapter focuses on the modeling of bulk and nanometric dielectric, and the final two chapters deal with specific intermetallic compounds and Heusler alloys. These chapters explain the fundamentals and present current research trends on magnetic skyrmions for technological innovation.

The topics presented here will be very helpful to students, researchers, academicians, and professionals. This book can also be used as a text for those who wish to engage in the physics of skyrmions in magnetic systems and related materials.

I would like to thank the authors who contributed their interesting and informative chapter for this book. Finally, I am pleased to thank all the members of the publishing house who have helped us with editing, revising, and reviewing this book and so much more.

Dipti Ranjan Sahu
Department of Natural and Applied Sciences,
Namibia University of Science and Technology,
Windhoek, Namibia

Chapter 1

Magnetic Skyrmions: Theory and Applications

Lalla Btissam Drissi, El Hassan Saidi, Mosto Bousmina and Omar Fassi-Fehri

Abstract

Magnetic skyrmions have been subject of growing interest in recent years for their very promising applications in spintronics, quantum computation and future low power information technology devices. In this book chapter, we use the field theory method and coherent spin state ideas to investigate the properties of magnetic solitons in spacetime while focussing on 2D and 3D skyrmions. We also study the case of a rigid skyrmion dissolved in a magnetic background induced by the spin-tronics; and derive the effective rigid skyrmion equation of motion. We examine as well the interaction between electrons and skyrmions; and comment on the modified Landau-Lifshitz-Gilbert equation. Other issues, including emergent electrodynamics and hot applications for next-generation high-density efficient information encoding, are also discussed.

Keywords: Geometric phases, magnetic monopoles and topology, soliton and holonomy, skyrmion dynamics and interactions, med-term future applications

1. Introduction

During the last two decades, the magnetic skyrmions and antiskyrmions have been subject to an increasing interest in connection with the topological phase of matter [1–4], the spin-tronics [5, 6] and quantum computing [7, 8]; as well as in the search for advanced applications such as racetrack memory, microwave oscillators and logic nanodevices making skyrmionic states very promising candidates for future low power information technology devices [9–12]. Initially proposed by T. Skyrme to describe hadrons in the theory of quantum chromodynamics [13], skyrmions have however been observed in other fields of physics, including quantum Hall systems [14, 15], Bose-Einstein condensates [16] and liquid crystals [17]. In quantum Hall (QH) ferromagnets for example [18, 19], due to the exchange interaction; the electron spins spontaneously form a fully polarized ferromagnet close to the integer filling factor $\nu \simeq 1$; slightly away, other electrons organize into an intricate spin configuration because of a competitive interplay between the Coulomb and Zeeman interactions [18]. Being quasiparticles, the skyrmions of the QH system condense into a crystalline form leading to the crystallization of the skyrmions [20–23]; thus opening an important window on promising applications.

In order to overcome the lack of a prototype of a skyrmion-based spintronic devices for a possible fabrication of nanodevices of data storage and logic technologies, intense research has been carried out during the last few years [24, 25]. In this regard, several alternative nano-objects have been identified to host stable skyrmions at room temperature. The first experimental observation of crystalline skyrmionic states was in a three-dimensional metallic ferromagnet MnSi with a B20 structure using small angle neutron scattering [26]. Then, real-space imaging of the skyrmion has been reported using Lorentz transmission electron microscopy in non-centrosymmetric magnetic compounds and in thin films with broken inversion symmetry, including monosilicides, monogermanides, and their alloys, like $Fe_{1-x}Co_xSi$ [27], FeGe [28], and MnGe [29].

One of the key parameters in the formation of these topologically protected non-collinear spin textures is the Dzyaloshinskii-Moriya Interaction (DMI) [30–32]. Originating from the strong spin-orbit coupling (SOC) at the interfaces, the DM exchange between atomic spins controls the size and stability of the induced skyrmions. Depending on the symmetry of the crystal structures and the skyrmion windings number, the internal spins within a single skyrmion envelop a sphere in different arrangements [33]. The in-plane component of the magnetization, in the Néel skyrmion, is always pointed in the radial direction [34], while it is oriented perpendicularly with respect to the position vector in the Bloch skyrmion [26]. Different from these two well-known types of skyrmions are skyrmions with mixed Bloch-Néel topological spin textures observed in Co/Pd multilayers [35]. Magnetic antiskyrmions, having a more complex boundary compared to the chiral magnetic boundaries of skyrmions, exist above room temperature in tetragonal Heusler materials [36]. Higher-order skyrmions should be stabilized in anisotropic frustrated magnet at zero temperature [37] as well as in itinerant magnets with zero magnetic field [38].

In the quest to miniaturize magnetic storage devices, reduction of material's dimensions as well as preservation of the stability of magnetic nano-scale domains are necessary. One possible route to achieve this goal is the formation of topological protected skyrmions in certain 2D magnetic materials. To induce magnetic order and tune DMIs in 2D crystal structures, their centrosymmetric should first be broken using some efficient ways such a (i) generate one-atom thick hybrids where atoms are mixed in an alternating manner [39–41], (ii) apply bias voltage or strain [42–44], (iii) insert adsorbents, impurities and defects [45–47]. In graphene-like materials, fluorine chemisorption is an exothermic adsorption that gives rise to stable 2D structures [48] and to long-range magnetism [49, 50]. In semi-fluorinated graphene, a strong Dzyaloshinskii-Moriya interaction has been predicted with the presence of ferromagnetic skyrmions [51]. The formation of a nanoskyrmion state in a Sn monolayer on a SiC(0001) surface has been reported on the basis of a generalized Hubbard model [52]. Strong DMI between the first nearest magnetic germanium neighbors in 2D semi-fluorinated germanene results in a potential anti-ferromagnetic skyrmion [53].

In this bookchapter, we use the coherent spin states approach and the field theory method (continuous limit of lattice magnetic models with DMI) to revisit some basic aspects and properties of magnetic solitons in spacetime while focusing on 1d kinks, 2d and 3d spatial skyrmions/antiskyrmions. We also study the case of a rigid skyrmion dissolved in a magnetic background induced by the electronic spins of magnetic atoms like Mn; and derive the effective rigid skyrmion equation of motion. In this regard, we describe the similarity between, on one hand, electrons in the electromagnetic background; and, on the other hand, rigid skyrmions bathing in a texture of magnetic moments. We also investigate the interaction between electrons and skyrmions as well as the effect of the spin transfer effect.

This bookchapter is organized as follows: In Section 2, we introduce some basic tools on quantum SU(2) spins and review useful aspects of their dynamics. In Section 3, we investigate the topological properties of kinks and 2d space solitons while describing in detail the underside of the topological structure of these low-dimensional solitons. In Section 4, we extend the construction to approach topological properties to 3d skyrmions. In Section 5, we study the dynamics of rigid skyrmions without and with dissipation; and in Section 6, we use emergent gauge potential fields to describe the effective dynamics of electrons interacting with the skyrmion in the presence of a spin transfer torque. We end this study by making comments and describing perspectives in the study of skyrmions.

2. Quantum SU(2) spin dynamics

In this section, we review some useful ingredients on the quantum SU(2) spin operator, its underlying algebra and its time evolution while focussing on the interesting spin 1/2 states, concerning electrons in materials; and on coherent spin states which are at the basis of the study of skyrmions/antiskyrmions. First, we introduce rapidly the SU(2) spin operator **S** and the implementation of time dependence. Then, we investigate the non dissipative dynamics of the spin by using semi-classical theory approach (coherent states). These tools can be also viewed as a first step towards the topological study of spin induced 1D, 2D and 3D solitons undertaken in next sections.

2.1 Quantum spin 1/2 operator and beyond

We begin by recalling that in non relativistic 3D quantum mechanics, the spin states $|S_z, S\rangle$ of spinfull particles are characterised by two half integers (S_z, S), a positive $S \geq 0$ and an S_z taking $2S+1$ values bounded as $-S \leq S_z \leq S$ with integral hoppings. For particles with spin 1/2 like electrons, one distinguishes two basis vector states $\left|\pm\frac{1}{2}, \frac{1}{2}\right\rangle$ that are eigenvalues of the scaled Pauli matrix $\frac{\hbar}{2}\sigma_z$ and the quadratic (Casimir) operator $\frac{\hbar^2}{4}\sum_{a=1}^{3}\sigma_a^2$, here the three $\frac{\hbar}{2}\sigma_a$ with $\sigma_a = \vec{\sigma}.\vec{e}_a$ are the three components of the spin 1/2 operator vector[1] $\vec{\sigma}$. From these ingredients, we learn that the average $< S_z, S|\frac{\hbar}{2}\sigma_z|S_z, S >= \hbar S_z$ (for short $\langle\frac{\hbar}{2}\sigma_z\rangle$) is carried by the z-direction since $S_z = \vec{S}.\vec{e}_z$ with $\vec{e}_z = (0,0,1)^T$. For generic values of the SU(2) spin S, the spin operator reads as $\hbar J_a$ where the three J_a's are $(2S+1)\times(2S+1)$ generators of the SU(2) group satisfying the usual commutation relations $[J_a, J_b] = i\varepsilon_{abc}J^c$ with ε_{abc} standing for the completely antisymmetric Levi-Civita tensor with non zero value $\varepsilon_{123} = 1$; its inverse is ε^{cba} with $\varepsilon^{123} = -1$. The time evolution of the spin $\frac{1}{2}$ operator $\hbar\frac{\sigma_a}{2}$ with dynamics governed by a stationary Hamiltonian operator $(dH/dt = 0)$ is given by the Heisenberg representation of quantum mechanics. In this non dissipative description, the time dependence of the spin $\frac{1}{2}$ operator $\hat{S}_a(t)$ (the hat is to distinguish the operator $\hat{S}_a$ from classical S_a) is given by

$$\hat{S}_a = e^{\frac{i}{\hbar}Ht}\left(\hbar\frac{\sigma_a}{2}\right)e^{-\frac{i}{\hbar}Ht} \tag{1}$$

where the Pauli matrices σ_a obey the usual commutation relations $[\sigma_a, \sigma_b] = 2i\varepsilon_{abc}\sigma^c$. For a generic value of the SU(2) spin S, the above relation extends as $\hat{S}_a = e^{\frac{i}{\hbar}Ht}(\hbar J_a)e^{-\frac{i}{\hbar}Ht}$. So, many relations for the spin 1/2 may be straightforwardly generalised for generic values S of the SU(2) spin. For example, for a spin value S_0, the $(2S_0+1)$ states are given by $\{|m, S_0\rangle\}$ and are labeled by $-S_0 \leq m \leq S_0$; one of these

[1] For convenience, we often refer to $\vec{\sigma}$, $\vec{e}_i$, $\vec{\sigma}.\vec{e}_i = \sigma_i$ respectively by bold symbols as $\boldsymbol{\sigma}$, $\mathbf{e}_i$, $\boldsymbol{\sigma}.\mathbf{e}_i = \sigma_i$.

states namely $|S_0, S_0\rangle$ is very special; it is commonly known as the highest weight state (HWS) as it corresponds to the biggest value $m = S_0$; from this state one can generate all other spin states $|m, S_0\rangle$; this feature will be used when describing coherent spin states. Because of the property $\sigma_a^2 = I$, the square $\hat{S}_a^2 = \frac{\hbar^2}{4} I$ is time independent; and then the time dynamics of $\hat{S}_a(t)$ is rotational in the sense that $\frac{d\hat{S}_a}{dt}$ is given by a commutator as follows $\frac{d\hat{S}_a}{dt} = \frac{i}{\hbar}\left(H\hat{S}_a - \hat{S}_a H\right)$. For the example where H is a linearly dependent function of $\hat{S}_a$ like for the Zeeman coupling, the Hamiltonian reads as $H_Z = \sum_a \omega^a \hat{S}_a$ (for short $\omega^a \hat{S}_a$) with the ω^a's are constants referring to the external source[2]; then the time evolution of $\hat{S}_a$ reads, after using the commutation relation $\left[\hat{S}_a, \hat{S}_b\right] = i\hbar\varepsilon_{abc}\hat{S}^c$, as follows

$$\frac{d\hat{S}_a}{dt} = \varepsilon_{abc}\omega^b \hat{S}^c \qquad \Leftrightarrow \qquad \frac{d\hat{\mathbf{S}}}{dt} = \omega \wedge \hat{\mathbf{S}} \tag{2}$$

where appears the Levi-Civita ε_{abc} which, as we will see throughout this study, turns out to play an important role in the study of topological field theory [54, 55] including solitons and skyrmions we are interested in here [56–59]. In this regards, notice that, along with this ε_{abc}, we will encounter another completely antisymmetric Levi-Civita tensor namely $\varepsilon_{\mu_1 \ldots \mu_D}$; it is also due to DM interaction which in lattice description is given by $\left(\vec{S}_{\mathbf{r}_{\mu_2}} \wedge \vec{S}_{\mathbf{r}_{\mu_1}}\right).\vec{d}_{\mu_3 \ldots \mu_{D-2}}\varepsilon^{\mu_1 \ldots \mu_D}$; and in continuous limit reads as $\varepsilon_{abc}S^b S^c_{\mu_1\mu_2} d^a_{\mu_3 \ldots \mu_{D-2}}\varepsilon^{\mu_1 \ldots \mu_D}$ where, for convenience, we have set $S^c_{\mu_1\mu_2} = \mathbf{e}_{\mu_1\mu_2}.\nabla S^c$ with $\mathbf{e}_{\mu_1\mu_2} = \mathbf{e}_{\mu_2} - \mathbf{e}_{\mu_1}$. To distinguish these two Levi-Civita tensors, we refer to ε_{abc} as the target space Levi-Civita with $SO(3)_{\text{target}}$ symmetry; and to $\varepsilon_{\mu_1 \ldots \mu_D}$ as the spacetime Levi-Civita with $SO(1, D-1)$ Lorentz symmetry containing as subsymmetry the usual space rotation group $SO(D-1)_{\text{space}}$. Notice also that for the case where the Hamiltonian $H\left(\hat{S}\right)$ is a general function of the spin, the vector ω^a is spin dependent and is given by the gradient $\frac{\partial H}{\partial \hat{S}_a}$.

2.2 Coherent spin states and semi-classical analysis

To deal with the semi-classical dynamics of $\hat{\mathbf{S}}(t)$ evolved by a Hamiltonian $H\left(\hat{\mathbf{S}}\right)$, we use the algebra $\left[\hat{S}_a, \hat{S}_b\right] = i\hbar\varepsilon_{abc}\hat{S}^c$ to think of the quantum spin in terms of a coherent spin state [60] described by a (semi) classical vector $\vec{S} = \hbar S\vec{n}$ (no hat) of the Euclidean $\mathbb{R}^3$; see the **Figure 1**(a). This "classical" 3-vector has an amplitude $\hbar S$ and a direction $\vec{n}$ related to a given unit vector $\vec{n}_0$ as $\vec{n} = R(\alpha, \beta, \gamma)\vec{n}_0$; and parameterised by α, β, γ. In the above relation, the $\vec{n}_0$ is thought of as the north direction of a 2-sphere $\mathbb{S}^2_{(\mathbf{n})}$ given by the canonical vector $(0, 0, 1)^T$; it is invariant under the proper rotation; i.e. $R_z(\gamma)\vec{n}_0 = \vec{n}_0$; and consequently the generic $\vec{n}$ is independent of γ; i.e.: $\vec{n} = R(\alpha, \beta)\vec{n}_0$. Recall that the 3×3 matrix $R(\alpha, \beta, \gamma)$ is an $SO(3)$ rotation [$SO(3) \sim SU(2)$] generating all other points of $\mathbb{S}^2_{(\mathbf{n})}$ parameterised by (α, β). In this regards, it is interesting to recall some useful properties that we list here after as three points: (1) the rotation matrix $R(\alpha, \beta, \gamma)$ can be factorised like

---

[2] For an electron with Zeeman field B^a, we have $\omega^a = -g\frac{q_e}{2m_e}B^a$ with $g = 2$ and $q_e = -e$.

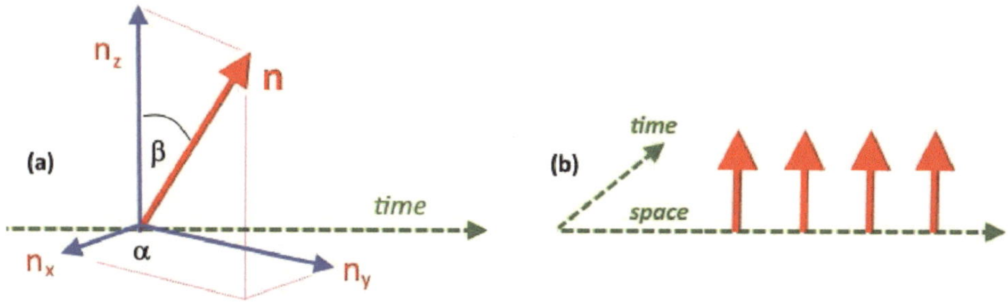

Figure 1.
(a) Components of the spin orientation n; its time dynamics in presence of a magnetic field is given by Larmor precession. (b) A configuration of several spins in spacetime.

$R_z(\alpha)R_y(\beta)R_z(\gamma)$ where each $R_a(\psi_a)$ is a rotation $e^{-i\psi_a J_a}$ around the a- axis with an angle ψ_a and generator J_a. (2) As the unit $\vec{n}_0$ is an eigen vector of $e^{-i\gamma J_z}$; it follows that $\vec{n}$ reduces to $e^{-i\alpha J_z}e^{-i\beta J_y}\vec{n}_0$; this generic vector obeys as well the constraint $\left|\vec{n}\right| \equiv |\mathbf{n}| = 1$ and is solved as follows.

$$\mathbf{n} = (\sin\beta\cos\alpha,\ \sin\beta\sin\alpha,\ \cos\beta) \tag{3}$$

with $0 \leq \alpha \leq 2\pi$ and $0 \leq \beta \leq \pi$; they parameterise the unit 2-sphere $\mathbb{S}^2_{(\mathbf{n})}$ which is isomorphic to $SU(2)/U(1)$; the missing angle γ parameterises a circle $\mathbb{S}^1_{(\mathbf{n})}$, isomorphic to $U(1)$, that is fibred over $\mathbb{S}^2_{(\mathbf{n})}$. (3) the coherent spin state representation gives a bridge between quantum spin operator and its classical description; it relies on thinking of the average $<\hat{\mathbf{S}}>$ in terms of the classical vector $\vec{S}_0 = \hbar S\vec{n}_0$ considered above ($\vec{S}_0 \leftrightarrow$ HWS $|S_0, S_0\rangle$). In this regards, recall that the $\hat{S}_a$ acts on classical 3-vectors V_b through its 3×3 matrix representation like $\left[\hat{S}_a, V_b\right] = -\hbar(J_c)_{ab}V^c$ with $(J_c)_{ab}$ given by $-i\varepsilon_{abc}$; these J_c's are precisely the generators of the $SU(2)$ matrix representation $R(\alpha,\beta,\gamma)$; by replacing V_b by the operator $\hat{S}_b$, one discovers the SU(2) spin algebra $\left[\hat{S}_a, \hat{S}_b\right] = i\hbar\varepsilon_{abc}\hat{S}^c$. Notice also that the classical spin vector $\vec{S} = \hbar S\vec{n}$ can be also put in correspondence with the usual magnetic moment $\vec{\mu} = -\gamma\vec{S}$ (with $\gamma = \frac{ge}{2m}$ the gyromagnetic ratio); thus leading to $\vec{\mu} = |\mu|\vec{n}$. So, the magnetization vector describes (up to a sign) a coherent spin state with amplitude $\hbar S\gamma$; and a (opposite) time dependent direction $\vec{n}(t)$ parameterizing the 2-sphere $\mathbb{S}^2_{(\mathbf{n})}$.

$$n_x^2(t) + n_y^2(t) + n_z^2(t) = 1 \Leftrightarrow \left|\vec{n}(t)\right| = 1 \tag{4}$$

For explicit calculations, this unit 2-sphere equation will be often expressed like $n^a n_a = 1$; this relation leads in turns to the property $n^a dn_a = 0$ (indicating that $\vec{n}$ and $d\vec{n}$ are normal vectors); by implementing time, the variation $\mathbf{n}.d\mathbf{n}$ gets mapped into $\mathbf{n}.\dot{\mathbf{n}} = 0$ teaching us that the velocity $\dot{\mathbf{n}}$ is carried by $\mathbf{u}$ and $\mathbf{v}$; two normal directions to $\mathbf{n}$ with components;

$$u_a = (\cos\beta\cos\alpha,\ \cos\beta\sin\alpha,\ -\sin\beta), \quad v_a = (-\sin\alpha,\ \cos\alpha, 0) \tag{5}$$

and from which we learn that $dn_a = u_a d\beta + v_a \sin\beta d\alpha$, $[(\mathbf{u}, \mathbf{v}, \mathbf{n})$ form an orthogonal vector triad). So, the dynamics of μ_a (and that of $-\vec{S}$) is brought to the dynamics of the unit n_a governed by a classical Hamiltonian $\mathrm{H}[n_a(\alpha,\beta)]$. The resulting time evolution is given by the so called Landau-Lifshitz (LL) equation [61]; it reads as $\frac{dn_a}{dt} = -\frac{\gamma}{|\mu|}\varepsilon_{abc}(\partial^b H)n^c$ with $\partial^b H = \frac{\partial H}{\partial n_b}$. By using the relations $d\beta = u^a dn_a$ and $\sin\beta d\alpha = v^a dn_a$ with $u_a = \frac{\partial n_a}{\partial \beta}$; and $v_a \sin\beta = \frac{\partial n_a}{\partial \alpha}$; as well as the expressions $\varepsilon_{abc}u^a n^c = v_b$ and $\varepsilon_{abc}v^a n^c = -u_b$, the above LL equation splits into two time evolution equations $\frac{d\beta}{dt} = -\gamma v_b(\partial^b H)$ and $\sin\beta\frac{d\alpha}{dt} = \gamma u_b(\partial^b H)$. These time evolutions can be also put into the form

$$\sin\beta\frac{d\beta}{dt} + \gamma\frac{\partial H}{\partial \alpha} = 0, \quad \sin\beta\frac{d\alpha}{dt} - \gamma\frac{\partial H}{\partial \beta} = 0 \tag{6}$$

and can be identified with the Euler- Lagrange equations following from the variation $\delta S = 0$ of an action $\mathcal{S} = \int L dt$. Here, the Lagrangian is related to the Hamiltonian like $L = L_B - H[n_a(\alpha,\beta)]$ where L_B is the Berry term [62] known to have the form <**n**|**ṅ**>; this relation can be compared with the well known Legendre transform p$\dot{q}$-H(q,p). For later interpretation, we scale this hamiltonian as $\hbar S\gamma H$ such that the spin lagrangian takes the form $L_{spin} = L_B - \hbar S\gamma H$. To determine L_B, we identify the Eq. (6) with the extremal variation $\delta S/\delta\beta = 0$ and $\delta S/\delta\alpha = 0$. Straightforward calculations leads to

$$L_B = -\hbar S(1 - \cos\beta)\frac{d\alpha}{dt} \tag{7}$$

showing that α and β form a conjugate pair. By substituting $\sin\beta\frac{d\alpha}{dt} = v^a\frac{dn_a}{dt}$ back into above L_B, we find that the Berry term has the form of Aharonov-Bohm coupling $L_{AB} = q_e A^a\frac{dn_a}{dt}$ with magnetic potential vector A^a given by $A_a = \frac{\hbar S}{q_e}\frac{(1-\cos\beta)}{\sin\beta}v_a$. However, this potential vector is suggestive as it has the same form as the potential vector $\mathbf{A}^{(monopole)} = \frac{\hbar S}{q_e r}\frac{(1-\cos\beta)}{\sin\beta}\mathbf{v}$ of a magnetic monopole. The curl of this potential is given by $\vec{\mathbf{B}} = q_m\frac{\vec{r}}{r^3}$ with magnetic charge $q_m = -\frac{\hbar S}{q_e}$ located at the centre of the 2-sphere; the flux Φ of this field through the unit sphere is then equal to $-4\pi\frac{\hbar S}{q_e}$; and reads as $-2S\Phi_0$ with a unit flux quanta $\Phi_0 = \frac{h}{q_e}$ as indicated by the value $S = 1/2$. So, because $2S = -n$ is an integer, it results that the flux is quantized as $\Phi = n\Phi_0$.

3. Magnetic solitons in lower dimensions

In previous section, we have considered the time dynamics of coherent spin states with amplitude $\hbar S$ and direction described by $\vec{n}(t)$ as depicted by the **Figure 1(a)**; this is a 3-vector having with no space coordinate dependence, $grad\vec{n} = 0$; and as such it can be interpreted as a $(1+0)D$ vector field; that is a vector belonging to $\mathbb{R}^{1,d}$ with $d = 0$ (no space direction). In this section, we first turn on 1d space coordinate x and promotes the old unit- direction $\vec{n}(t)$ to a $(1+1)$D field $\vec{n}(t,x)$. After that, we turn on two space directions (x,y); thus leading to $(1+2)$D field $\vec{n}(t,x,y)$; a picture is depicted by the **Figure 1(b)**. To deal with the dynamics of these local fields and their topological properties, we use the field theory method while focussing on particular solitons; namely the 1d kinks and the 2d skyrmions. In this extension, one encounters two types of spaces: (1) the target space $\mathbb{R}^3_{\mathbf{n}}$ parameterised by $n_a = (n_1, n_2, n_3)$ with Euclidian metric δ_{ab} and topological Levi-Civita ε_{abc}. (2) the spacetime $\mathbb{R}^{1,1}_{\xi}$ parameterised by $\xi^\mu = (t, x)$, concerning the 1d kink evolution; and the spacetime $\mathbb{R}^{1,2}_{\xi}$ parameterised by $\xi^\mu = (t, x, y)$, regarding the 2d skyrmions dynamics. As we have two kinds of evolutions; time and space; we denote the time variable by $\xi^0 = t$; and the space coordinates by $\xi^i = (x, y)$. Moreover, the homologue of the tensors δ_{ab} and ε_{abc} are respectively given by the usual Lorentzian spacetime metric $g_{\mu\nu}$, with signature like $g_{\mu\nu}\xi^\mu\xi^\nu = x^2 + y^2 - t^2$, and the spacetime Levi-Civita $\varepsilon_{\mu\nu\rho}$ with $\varepsilon_{012} = 1$.

3.1 One space dimensional solitons

In $(1+1)$D spacetime, the local coordinates parameterising $\mathbb{R}^{1,1}_{(\xi)}$ are given by $\xi^\mu = (t, x)$; so the metric is restricted to $g_{\mu\nu}\xi^\mu\xi^\nu = x^2 - t^2$. The field variable $n^a(\xi)$ has in general three components (n_1, n_2, n_3) as described previously; but in what follows, we will simplify a little bit the picture by setting $n_3 = 0$; thus leading to a magnetic 1d soliton with two component field variable $\mathbf{n} = (n_1, n_2)$ satisfying the constraint equation $\mathbf{n}.\mathbf{n} = 1$ at each point of spacetime. As this constraint relation plays an important role in the construction, it is interesting to express it as $n_a n^a = 1$. Before describing the topological properties of one space dimensional solitons (kinks), we think it interesting to begin by giving first some useful features; in particular the three following ones. (1) The constraint $(n_1)^2 + (n_2)^2 = 1$ is invariant $SO(2)_{\mathbf{n}}$ rotations acting as $n'^a = \mathcal{R}^a_b n^b$ with orthogonal rotation matrix

$$\mathcal{R}^a_b = \begin{pmatrix} \cos\psi & \sin\psi \\ -\sin\psi & \cos\psi \end{pmatrix}, \qquad \mathcal{R}^T\mathcal{R} = I \tag{8}$$

The constraint $n_a n^a = 1$ can be also presented like $\overline{N}N = 1$ with N standing for the complex field $n_1 + in_2$ that reads also like $e^{i\alpha}$. In this complex notation, the symmetry of the constraint is given by the phase change acting as $N \to UN$ with $U = e^{i\psi}$ and corresponding to the shift $\alpha \to \alpha + \psi$. Moreover the correspondence $(n_1, n_2) \leftrightarrow n_1 + in_2$ describes precisely the well known isomorphisms $SO(2) \sim U(1) \sim \mathbb{S}^1_{(\mathbf{n})}$ where $\mathbb{S}^1_{(\mathbf{n})}$ is a circle; it is precisely the equatorial circle of the 2-sphere $\mathbb{S}^2_{(\mathbf{n})}$ considered in previous section. (2) As for Eq. (5), the constraint $n_a n^a = 1$ leads to $n_a dn^a = 0$; and so describes a rotational movement encoded in the relation $dn^a = \varepsilon^{ab} n_b$ where ε^{ab} is the standard 2D antisymmetric tensor with $\varepsilon^{21} = \varepsilon_{12} = 1$; this ε_{ab} is related to the previous 3D Levi-Civita like ε_{zab}. Notice also that the constraint $n_a n^a = 1$ implies moreover that $dn_2 = -\frac{n_n}{n_2} dn_1$; and consequently the area $dn_1 \wedge dn_2$, to be encountered later on, vanishes identically. In this regards, recall that we have the following transformation

$$dn_1 \wedge dn_2 = \mathfrak{J} dt \wedge dx, \quad \mathfrak{J} = \varepsilon^{\mu\nu}\partial_\mu n_1 \partial_\nu n_2 \tag{9}$$

where $\varepsilon^{\mu\nu}$ is the antisymmetric tensor in 1+1 spacetime, and $\mathfrak{J}$ is the Jacobian of the transformation $(t, x) \to (n_1, n_2)$. (3) The condition $n_a n^a = 1$ can be dealt in two manners; either by inserting it by help of a Lagrange multiplier; or by solving it in term of a free angular variable like $n_a = (\cos\alpha,\ \sin\alpha)$ from which we deduce the normal direction $u^a = \frac{dn^a}{d\alpha}$ reading as $u_a = (-\sin\alpha,\ \cos\alpha)$. In term of the complex field; we have $N = e^{i\alpha}$ and $\overline{N}dN = id\alpha$. Though interesting, the second way of doing hides an important property in which we are interested in here namely the non linear dynamics and the topological symmetry.

3.1.1 Constrained dynamics

The classical spacetime dynamics of $n^a(\xi)$ is described by a field action $\mathcal{S} = \int dtL$ with Lagrangian $L = \int dx\mathcal{L}$ and density $\mathcal{L}$; this field density is given by $-\frac{1}{2}(\partial_\mu n_a)(\partial^\mu n^a) - V(n) - \Lambda(n^a n_a - 1)$ with $\partial_\mu = \frac{\partial}{\partial \xi^\mu}$; it reads in terms of the Hamiltonian density as follows

$$\mathcal{L} = \pi^a \dot{n}_a - \mathcal{H} \tag{10}$$

where $\pi^a = \frac{\partial \mathcal{L}}{\partial \dot{n}_a}$. In the above Lagrangian density, the auxiliary field $\Lambda(\xi)$ (no Kinetic term) is a Lagrange multiplier carrying the constraint relation $n_a n^a = 1$. The $V(n)$ is a potential energy density which play an important role for describing 1d kinks with finite size. Notice also that the variation $\frac{\delta S}{\delta \Lambda} = 0$ gives precisely the constraint $n_a n^a = 1$ while the $\frac{\delta S}{\delta n^a} = 0$ gives the spacetime dynamics of n^a described by the spacetime equation $\partial_\mu \partial^\mu n^a - \frac{\partial V}{\partial n^a} - \Lambda n^a = 0$. By substituting $n_a = (\cos\alpha, \sin\alpha)$, we obtain $\mathcal{L} = -\frac{1}{2}(\partial_\mu \alpha)(\partial^\mu \alpha) - V(\alpha)$. If setting $V(\alpha) = 0$, we end up with the free field equation $\partial_\mu \partial^\mu \alpha = 0$ that expands like $(\partial_x^2 - \partial_t^2)\alpha = 0$; it is invariant under spacetime translations with conserved current symmetry $\partial^\mu T_{\mu\nu} = 0$ with $T_{\mu\nu}$ standing for the energy momentum tensor given by the 2×2 symmetric matrix $\partial_\mu \alpha \partial_\nu \alpha + g_{\mu\nu}\mathcal{L}$. The energy density T_{00} is given by $\frac{1}{2}(\partial_t \alpha)^2 + \frac{1}{2}(\partial_x \alpha)^2$ and the momentum density T_{10} reads as $\partial_x \varphi \partial_x \varphi$. Focussing on T_{00}, the conserved energy E reads then as follows

$$E = \frac{1}{2}\int_{-\infty}^{+\infty} dx \left[(\partial_t \alpha)^2 + (\partial_x \alpha)^2\right] \geq 0 \tag{11}$$

with minimum corresponding to constant field ($\alpha = cte$). Notice that general solutions of $\partial_\mu \partial^\mu \alpha = 0$ are given by arbitrary functions $f(x \pm t)$; they include oscillating and non oscillating functions. A typical non vibrating solution that is interesting for the present study is the solitonic solution given (up to a constant c) by the following expression

$$\varphi(t, x) = \pi \tanh\left(\frac{x+t}{\lambda}\right) \tag{12}$$

where λ is a positive parameter representing the width where the soliton $\alpha(t, x)$ acquires a significant variation. Notice that for a given t, the field varies from $\alpha(t, -\infty) = -\pi$ to $\alpha(t, +\infty) = \pi$ regardless the value of λ. These limits are related to each other by a period 2π.

3.1.2 Topological current and charge

To start, notice that as far as conserved symmetries of (10) are concerned, there exists an exotic invariance generated by a conserved $J_\mu(t, x)$ going beyond the spacetime translations generated by the energy momentum tensor $T_{\mu\nu}$. The conserved spacetime current $J_\mu = (J_0, J_1)$ of this exotic symmetry can be introduced in two different, but equivalent, manners; either by using the free degree of freedom α; or by working with the constrained field n^a. In the first way, we think of the charge density J_0 like $\frac{1}{2\pi}\partial_1\alpha$ and of the current density as $J_1 = -\frac{1}{2\pi}\partial_0\alpha$. This conserved current is a topological $(1+1)D$ spacetime vector J_μ that is manifestly conserved; this feature follows from the relation between J_μ and the antisymmetric $\varepsilon_{\mu\nu}$ as follows [57],

$$J_\mu = \frac{1}{2\pi}\varepsilon_{\mu\nu}\partial^\nu\alpha \tag{13}$$

Because of the $\varepsilon_{\mu\nu}$; the continuity relation $\partial^\mu J_\mu = \frac{1}{2\pi}\varepsilon_{\mu\nu}\partial^\mu\partial^\nu\alpha$ vanishes identically due to the antisymmetry property of $\varepsilon_{\mu\nu}$. The particularity of the above conserved J_μ is its topological nature; it is due to the constraint $n_a n^a = 1$ without recourse to the solution $n_a = (\cos\alpha, \sin\alpha)$. Indeed, Eq. (13) can be derived by computing the Jacobian $\mathfrak{J} = \det\left(\frac{\partial n^a}{\partial \xi^\mu}\right)$ of the mapping from the 2d spacetime coordinates (t, x) to the target space fields (n_1, n_2). Recall that the spacetime area $dt \wedge dx$ can be written in terms of $\varepsilon_{\mu\nu}$ like $\frac{1}{2}\varepsilon_{\mu\nu}d\xi^\mu \wedge d\xi^\nu$ and, similarly, the target space area $dn_1 \wedge dn_2$ can be expressed in terms of as ε_{ab} follows $\frac{1}{2}\varepsilon_{ab}dn^a \wedge dn^b$. The Jacobian $\mathfrak{J}$ is precisely given by (9); and can be presented into a covariant form like $\mathfrak{J} = \frac{1}{2}\varepsilon^{\mu\nu}\partial_\mu n^a \partial_\nu n^b \varepsilon_{ab}$. This expression of the Jacobian $\mathfrak{J}$ captures important informations; in particular the three following ones. (1) It can be expressed as a total divergence like $\partial_\mu(\pi J^\mu)$ with spacetime vector

$$J^\mu = \frac{1}{2\pi}\varepsilon^{\mu\nu}n^a\partial_\nu n^b\varepsilon_{ab} \tag{14}$$

and where $\frac{1}{\pi}$ is a normalisation; it is introduced for the interpretation of the topological charged as just the usual winding number of the circle [encoded in the homotopy group relation $\pi_1(\mathbb{S}^1) = \mathbb{Z}$]. (2) Because of the constraint $dn_2 = -\frac{n_1}{n_2}dn_1$ following from $n_a n^a = 1$, the Jacobian $\mathfrak{J}$ vanishes identically; thus leading to the conservation law $\partial_\mu J^\mu = 0$; i.e. $\mathfrak{J} = 0$ and then $\partial_\mu J^\mu = 0$. (3) The conserved charge Q associated with the topological current is given by $\int_{-\infty}^{+\infty} dx J^0(t, x)$; it is time independent despite the apparent t- variable in the integral $(dQ/dt = 0)$. By using (13), this charge reads also as $\frac{1}{2\pi}\int_{-\infty}^{+\infty} dx\partial_x\alpha(t, x)$ and after integration leads to

$$Q = \frac{1}{2\pi}[\alpha(t, \infty) - \alpha(t, -\infty)] \tag{15}$$

Moreover, seen that $\alpha(t, \infty)$ is an angular variable parameterising $\mathbb{S}^1_{\mathbf{n}}$; it may be subject to a boundary condition like for instance the periodic $\alpha(t, \infty) = \alpha(t, -\infty) + 2\pi N$ with N an integer; this leads to an integral topological charge $Q = N$ interpreted as the winding number of the circle. In this regards, notice that: (i) the winding interpretation can be justified by observing that under compactification of the space variable x, the infinite space line $\mathbb{R}_x =]-\infty, +\infty[$ gets mapped into a circle $\mathbb{S}^1_{(x)}$ with angular coordinate $-\pi \le \varphi \le \pi$; so, the integral $\frac{1}{2\pi}\int_{-\infty}^{+\infty} dx\partial_x\alpha(t, x)$ gets replaced by $\frac{1}{2\pi}\int_{-\pi}^{+\pi} d\varphi\frac{\partial\alpha}{\partial\varphi}$; and then the mapping $\alpha_t : \varphi \to \alpha(t, \varphi)$ is a mapping between two circles namely $\mathbb{S}^1_{(x)} \to \mathbb{S}^1_{(\mathbf{n})}$; the field $\alpha(t, \varphi)$ then describes a soliton (one space extended object) wrapping the circle $\mathbb{S}^1_{(x)}$ N times; this propery is captured by $\pi_{\mathbb{S}^1_{(x)}}\left(\mathbb{S}^1_{(\mathbf{n})}\right) = \mathbb{Z}$, a homotopy group property [63]. (ii) The charge Q is independent of the Lagrangian of the system as it follows completely from the field constraint

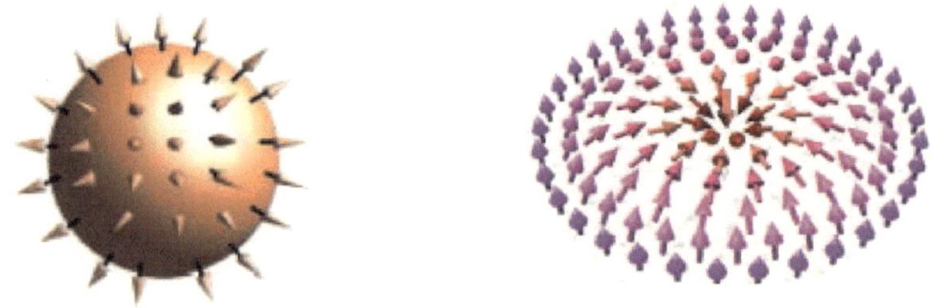

Figure 2.
On left: a spin configuration with $n_1^2 + n_2^2 + n_3^2 = 1$ dispatched on a 2-sphere. On right: a two space dimensional magnetic skyrmion given by the stereographic projection of $\mathbb{S}^2$ to plane.

without any reference to the field action. (*iii*) Under a scale transformation $\xi' = \xi/\lambda$ with a scaling parameter $\lambda > 0$, the topological charge of the field (12) is invariant; but its total energy (11) get scaled as follows

$$Q' = Q, \qquad E' = \frac{1}{\lambda}E \tag{16}$$

This energy transformation shows that stable solitons with minimal energy correspond to $\lambda \to \infty$; and then to a trivial soliton spreading along the real axis. However, one can have non trivial solitonic configurations that are topologically protected and energetically stable with non diverging λ. This can done by turning on an appropriate potential energy density $V(\mathbf{n})$ in Eq. (10). An example of such potential is the one given by $\frac{g}{8}\left(n_1^4 + n_2^4 - 1\right)$, with positive $g = M^2$, breaking $SO(2)_{\mathbf{n}}$; by using the constraint $n_1^2 + n_2^2 = 1$, it can be put $\frac{g}{4}n_1^2 n_2^2$. In terms of the angular field α, it reads as $V(\alpha) = \frac{g}{16}(1 - \cos 4\alpha)$ leading to the well known sine-Gordon Eq. [64, 65] namely $\partial_\mu \partial^\mu \alpha - \frac{g}{4}\sin 4\alpha = 0$ with the symmetry property $\alpha \to \alpha + \frac{\pi}{2}$. So, the solitonic solution is periodic with period $\frac{\pi}{2}$; that is the quarter of the old 2π period of the free field case. For static field $\alpha(x)$, the sine Gordon equation reduces to $\frac{d^2\alpha}{dx^2} - \frac{M^2}{4}\sin 4\alpha = 0$; its solution for $M > 0$ is given by $\arctan\left[\exp Mx\right]$ representing a sine- Gordon field evolving from 0 to $\frac{\pi}{2}$ and describing a kink with topological charge $Q = \frac{1}{4}$. For $M < 0$, the soliton is an anti-kink evolving from $\frac{\pi}{2}$ to 0 with charge $Q = -\frac{1}{4}$. Time dependent solutions can be obtained by help of boost transformations $x \to \frac{x \pm vt}{\sqrt{1-v^2}}$.

3.2 Skyrmions in 2d space dimensions

In this subsection, we investigate the topological properties of 2d Skyrmions by extending the field theory study we have done above for 1d kinks to two space dimensions. For that, we proceed as follows: First, we turn on the component n_3 so that the skyrmion field $\mathbf{n}$ is a real 3-vector with three components (n_1, n_2, n_3) constrained as in Eqs. (4) and (5); see **Figure 2**. Second, here we have $\mathbf{n} = \mathbf{n}(t, x, y)$; that is a 3-component field living in the $(2 + 1)$ space time with Lorentzian metric and coordinates $\xi^\mu = (t, x, y)$. This means that $d\mathbf{n} = (\partial_\mu \mathbf{n})d\xi^\mu$; explicitly $d\mathbf{n} = \frac{\partial \mathbf{n}}{\partial t}dt + \frac{\partial \mathbf{n}}{\partial x}dx + \frac{\partial \mathbf{n}}{\partial y}dy$.

3.2.1 Dzyaloshinskii-Moriya potential

The field action $\mathcal{S}_{3D} = \int dt L_{3D}$ describing the space time dynamics of $\mathbf{n}(t, x, y)$ has the same structure as Eq. (10); except that here the Lagrangian L_{3D} involves two space variable like $\int dxdy\mathcal{L}_{3D}$ and the density $\mathcal{L}_{3D} = -\frac{1}{2}\left(\partial_\mu \mathbf{n}\right)^2 - V(\mathbf{n}) - \Lambda(\mathbf{n}.\mathbf{n} - 1)$; this is a function of the constrained 3-vector $\mathbf{n}$ and its space time gradient $\partial_\mu \mathbf{n}$; it reads in term of the Hamiltonian density as follows.

$$\mathcal{L}_{3D} = \pi.\dot{\mathbf{n}} - \mathcal{H}_{3D}(\mathbf{n}) \tag{17}$$

In this expression, the $\mathcal{H}_{3D}(\mathbf{n})$ is the continuous limit of a lattice Hamiltonian $\mathrm{H}_{latt}\left(\left[n^a\left(\mathbf{r}_\mu\right)\right]\right.$ involving, amongst others, the Heisenberg term, the Dyaloshinskii-Moriya (DM) interaction and the Zeeman coupling. The $V(\mathbf{n})$ in the first expression of $\mathcal{L}_{3D}$ is the scalar potential energy density; it models the continuous limit of the interactions that include the DM and Zeeman ones [see Eq. (1.22) for its explicit relation]. The field $\Lambda(\xi)$ is an auxiliary 3D spacetime field; it is a Lagrange multiplier that carries the constraint $\mathbf{n}.\mathbf{n} = 1$ which plays the same role as in subSection 3.1. By variating this action with respect to the fields $\mathbf{n}$ and Λ; we get from $\frac{\delta S_{3D}}{\delta \Lambda} = 0$ precisely the field constraint $\mathbf{n}.\mathbf{n} = 1$; and from $\frac{\delta S_{3D}}{\delta \mathbf{n}} = 0$ the following Euler–Lagrange equation $\mathrm{W}\mathbf{n} = \frac{\partial V}{\partial \mathbf{n}} + \Lambda \mathbf{n}$. For later use, we express this field equation like

$$\partial^\mu \partial_\mu n_a = \frac{\partial V}{\partial n^a} + \Lambda n_a \tag{18}$$

The interest into this (18) is twice; first it can be put into the equivalent form $\partial^\mu \partial_\mu n_a = \varepsilon_{abc}\mathcal{D}^b n^c$ where $\mathcal{D}^b$ is an operator acting on n^c to be derived later on [see Eq. (22) given below]; and second, it can be used to give the relation between the scalar potential and the operator $\mathcal{D}^b$. To that purpose, we start by noticing that there are two manners to deal with the field constraint $n^a n_a = 1$; either by using the Lagrange multiplier Λ; or by solving it in terms of two angular field variables as given by Eq. (5). In the second case, we have the triad $n_a = (\sin\beta\cos\alpha,\ \sin\beta\sin\alpha,\ \cos\beta)$ and

$$u_a = (\cos\beta\cos\alpha,\ \cos\beta\sin\alpha, -\sin\beta) \quad , \quad v_a = (-\sin\alpha,\ \cos\alpha, 0) \tag{19}$$

but now $\beta = \beta(t,x,y)$ and $\alpha = \alpha(t,x,y)$ with $0 \leq \beta \leq \pi$ and $0 \leq \alpha \leq 2\pi$. Notice also that the variation of the filed constraint leads to $n_a dn^a = 0$ teaching us interesting informations, in particular the two following useful ones. (1) the movement of n_a in the target space is a rotational movement; and so can be expressed like

$$dn_a = \varepsilon_{abc}\omega^b n^c \quad \Leftrightarrow \quad d\mathbf{n} = \omega \wedge \mathbf{n} \quad \Leftrightarrow \quad \omega \sim \mathbf{n} \wedge d\mathbf{n} \tag{20}$$

where the 1-form ω^b is the rotation vector to be derived below. By substituting (20) back into $n^a dn_a$, we obtain $\varepsilon_{bca}\omega^b n^c n^a$ which vanishes identically due to the property $\varepsilon_{bca}n^c n^a = 0$. (2) Having two degrees of freedom α and β, we can expand the differential dn_a like $u_a d\beta + v_a \sin\alpha d\alpha$ with the two vector fields $u_a = \frac{\partial n_a}{\partial \beta}$ and $v_a = \frac{\partial n_a}{\partial \alpha}$ as given above. Notice that the three unit fields $(\mathbf{n}, \mathbf{u}, \mathbf{v})$ plays an important role in this study; they form a vector basis of the field space; they obey the usual cross products namely $\mathbf{n} = \mathbf{u} \wedge \mathbf{v}$ and its homologue which given by cyclic permutations; for example,

$$u_a = \varepsilon_{abc}v^b n^c, \quad v_a = -\varepsilon_{abc}u^b n^c \tag{21}$$

Putting these Eq. (21) back into the expansion of dn_a in terms of $d\alpha$, $d\beta$; and comparing with Eq. (20), we end up with the explicit expression of the 1-form angular "speed" vector ω^b; it reads as follows $\omega^b = v^b d\beta - u^b \sin\alpha d\alpha$. Notice that by using the space time coordinates ξ, we can also express Eq. (20) like $\partial_\mu n_a = \varepsilon_{abc}\omega_\mu^b n^c$ with ω_μ^b given by $v^b(\partial_\mu\beta) - u^b \sin\alpha(\partial_\mu\alpha)$. From this expression, we can compute the Laplacian $\partial^\mu\partial_\mu n_a$; which, by using the above relations; is equal to $\varepsilon_{abc}\partial^\mu\left(\omega_\mu^b n^c\right)$ reading explicitly as $\varepsilon_{abc}\left[\left(\partial^\mu\omega_\mu^b\right)n^c + \omega_\mu^b(\partial^\mu n^c)\right]$ or equivalently like $\partial^\mu\partial_\mu n_a =$ $\varepsilon_{abc}\mathcal{D}^b n^c$ with operator $\mathcal{D}^b = \omega_\mu^b\partial^\mu + \left(\partial^\mu\omega_\mu^b\right)$. Notice that the above operator has an interesting geometric interpretation; by factorising ω_μ^b, we can put it in the form $\omega_\mu^d(\mathcal{D}^\mu)_d^b$ where $(\mathcal{D}^\mu)_d^b$ appears as a gauge covariant derivative $(\mathcal{D}^\mu)_d^b = \delta_d^b\partial^\mu + (A^\mu)_d^b$ with a non trivial gauge potential $(A^\mu)_d^b$ given by $\omega_d^\mu(\partial^\nu\omega_\nu^b)$. Comparing with (18) with $\partial^\mu\partial_\mu n_a = \varepsilon_{abc}\mathcal{D}^b n^c$, we obtain $\frac{\partial V}{\partial n^a} = \varepsilon_{abc}\mathcal{D}^b n^c - \Lambda n_a$; and then a scalar potential energy V given by $\int \varepsilon_{abc}(dn^a\mathcal{D}^b n^c) - \Lambda\int n_a dn^a$. The second term in this relation vanishes identically because $n_a dn^a = 0$; thus reducing to

$$V = \int \varepsilon_{abc}\left(dn^a\mathcal{D}^b n^c\right) \tag{22}$$

containing $\varepsilon_{abc}(n^a\mathcal{D}^b n^c)$ as a sub-term. In the end of this analysis, let us compare this sub-term with the $\varepsilon_{abc}n^b n^c_{\mu_1\mu_2}\Delta^{a\mu_1\mu_2}$ with $\Delta^{a\mu_1\mu_2} = d^a_{\mu_3\cdots\mu_{D-2}}\varepsilon^{\mu_1\cdots\mu_D}$ giving the general structure of the DM coupling (see end of subSection 2.1). For $(1+2)$D spacetime, the general structure of DM interaction reads $\varepsilon_{abc}d_0^a(n^b\nabla n^c).\mathbf{e}_{\mu\nu}\varepsilon^{0\mu\nu}$; by setting $\mathbf{e}^0 = \mathbf{e}_{\mu\nu}\varepsilon^{0\mu\nu}$ and $\mathbf{d}^a = d_0^a\mathbf{e}^0$ as well as $D^a = \mathbf{d}^a.\nabla$, one brings it to the form $\varepsilon_{abc}(n^a D^b n^c)$ which is the same as the one following from (22).

3.2.2 From kinks to 2d Skyrmions

Here, we study the topological properties of the 2d Skyrmion with dynamics governed by the Lagrangian density (17). From the expression of the $(1+1)$D topological current $(J^\mu)_{2D}$ discussed in subSection 2.1, which reads as $\frac{1}{2\pi}\varepsilon^{\mu\nu}n^a\partial_\nu n^b\varepsilon_{ab}$, one can wonder the structure of the $(1+2)$D topological current $(J^\mu)_{3D}$ that is associated with the 2d Skyrmion described by the 3-vector field $n_a(\xi)$. It is given by

$$(J^\mu)_{3D} = \frac{1}{8\pi}\varepsilon^{\mu\nu\rho}n^a\partial_\nu n^b\partial_\nu n^c\varepsilon_{abc} \tag{23}$$

where ε_{abc} is as before and where $\varepsilon^{\mu\nu\rho}$ is the completely antisymmetric Levi-Civita tensor in the $(1+2)$D spacetime. The divergence $\partial_\mu(J^\mu)_{3D}$ of the above spacetime vector vanishes identically; it has two remarkable properties that we want to comment before proceeding. (1) The $\partial_\mu(J^\mu)_{3D}$ is nothing but the determinant of the 3×3 Jacobian matrix $\frac{\partial n^a}{\partial\xi^\mu}$ relating the three field variables n^a to the three spacetime coordinates ξ^μ; this Jacobian $\det\left(\frac{\partial n^a}{\partial\xi^\mu}\right)$ is generally given by $\frac{1}{3!}\varepsilon^{\mu\nu\rho}\partial_\mu n^a\partial_\nu n^b\partial_\nu n^b\varepsilon_{abc}$; it maps the spacetime volume $d^3\xi = dt\wedge dx\wedge dy$ into the target space volume $d^3\mathbf{n} = dn^1\wedge dn^2\wedge dn^3$. In this regards, recall that these two 3D volumes can be expressed in covariant manners by using the completely antisymmetric tensors $\varepsilon_{\mu\nu\rho}$ and ε_{abc} introduced earlier; and as noticed before play a central role in topology. The target space volume $d^3\mathbf{n}$ can be expressed like $\frac{1}{3!}\varepsilon_{abc}dn^a\wedge dn^b\wedge dn^c$; and a similar relation can be also written down for the spacetime volume $d^3\xi$. Notice also that by substituting the differentials dn^a by their expansions $\left(\frac{\partial n^a}{\partial \xi^\mu}\right) d\xi^\mu$; and putting back into $d^3\mathbf{n}$, we obtain the relation $d^3\mathbf{n} = \mathfrak{J}_{3D} d^3\xi$ where $\mathfrak{J}_{3D}$ is precisely the Jacobian $\det\left(\frac{\partial n^a}{\partial \xi^\mu}\right)$. (2) The conservation law $\partial_\mu (J^\mu)_{3D} = 0$ has a geometric origin; it follows from the field constraint relation $n_1^2 + n_2^2 + n_3^2 = 1$ degenerating the volume of the 3D target space down to a surface. This constraint relation describes a unit 2-sphere $\mathbb{S}^2_{(\mathbf{n})}$; and so a vanishing volume $d^3\mathbf{n}|_{\mathbb{S}^2_{(\mathbf{n})}} = 0$; thus leading to $\mathfrak{J}_{3D} = 0$ and then to the above continuity equation. Having the explicit expression (23) of the topological current J^μ in terms of the magnetic texture field $n(\xi)$, we turn to determine the associated topological charge $Q = \int dxdyJ^0$ with charge density J^0 given by $\frac{1}{8\pi}\varepsilon_{abc}\varepsilon^{0ij}(\partial_i n^b \partial_j n^c) n^a$. Substituting $\varepsilon^{0ij} dx \wedge dy$ by $d\xi^i \wedge d\xi^j$, we have $J^0 dx \wedge dy = \frac{1}{8\pi}\varepsilon_{abc} n^a \left(dn^b \wedge dn^c\right)$. Moreover using the differentials $dn^b = u^b d\beta + v^b \sin\alpha d\alpha$, we can calculate the area $dn^b \wedge dn^c$ in terms of the angles α and β; we find $2n_a(\sin\alpha) d\beta \wedge d\alpha$ where we have used $\varepsilon_{abc}\left(u^b v^c - u^c v^b\right) = 2n_a$. So, the topological charge Q reads as $\frac{1}{4\pi}\int_{\mathbb{S}^2_{\mathbf{n}}} (\sin\beta) d\alpha d\beta$ which is equal to 1. In fact this value is just the unit charge; the general value is an integer $Q = N$ with N being the winding number $\pi_2(\mathbb{S}^2_{\mathbf{n}})$; see below. Notice that J^0 can be also presented like

$$J^0 = \frac{\varepsilon_{abc}}{8\pi} n^a \left(\frac{\partial n^b}{\partial x}\frac{\partial n^c}{\partial y} - \frac{\partial n^b}{\partial y}\frac{\partial n^c}{\partial x}\right) \tag{24}$$

Replacing n_a by their expression in terms of the angles $(\sin\beta\cos\alpha,\ \sin\beta\sin\alpha,\ \cos\beta)$, we can bring the above charge density J^0 into two equivalent relations; first into the form like $\frac{\sin\beta}{4\pi}\left(\frac{\partial\beta}{\partial x}\frac{\partial\alpha}{\partial y} - \frac{\partial\beta}{\partial y}\frac{\partial\beta}{\partial x}\right)$; and second as $\frac{1}{4\pi}\frac{\partial[\alpha,\ \cos\beta]}{\partial[x,y]}$ which is nothing but the Jacobian of the transformation from the (x,y) space to the unit 2-sphere with angular variables (α,β). The explicit expression of (n_1, n_2, n_3) in terms of the (x,y) space variables is given by

$$n_1 = \frac{2x}{x^2+y^2+1} \quad , \quad n_2 = \frac{2y}{x^2+y^2+1} \quad , \quad n_3 = \frac{x^2+y^2-1}{x^2+y^2+1} \tag{25}$$

but this is nothing but the stereographic projection of the 2-sphere $\mathbb{S}^2_\xi$ on the real plane. So, the field n^a defines a mapping between $\mathbb{S}^2_\xi$ towards $\mathbb{S}^2_{\mathbf{n}}$ with topological charge given by the winding number $\mathbb{S}^2_{\mathbf{n}}$ around $\mathbb{S}^2_{\mathbf{n}}$; this corresponds just to the homotopy property $\pi_2(\mathbb{S}^2_{\mathbf{n}}) = N$.

4. Three dimensional magnetic skyrmions

In this section, we study the dynamics of the 3d skyrmion and its topological properties both in target space $\mathbb{R}^4_{\boldsymbol{n}}$ (with euclidian metric δ_{AB}) and in 4D spacetime $\mathbb{R}^{1,3}_\xi$ parameterised by $\xi^\mu = (t,x,y,z)$ (with Lorentzian metric $g_{\mu\nu}$). The spacetime dynamics of the 3d skyrmion is described by a four component field $\boldsymbol{n}_A(\xi)$ obeying a constraint relation $f(\boldsymbol{n}) = 1$; here the $f(\boldsymbol{n})$ is given by the quadratic form $\boldsymbol{n}_A \boldsymbol{n}^A$ invariant under SO(4) transformations isomorphic to SU(2)×SU(2). The structure of the topological current of the 3d skyrmion is encoded in two types of Levi-Civita tensors namely the target space ε_{ABCD} and the spacetime $\varepsilon_{\mu\nu\rho\tau}$ extending their homologue concerning the kinks and 2d skyrmions.

4.1 From 2d skyrmion to 3d homologue

As for the 1d and 2d solitons considered previous section, the spacetime dynamics of the 3d skyrmion in $\mathbb{R}^{1,3}$ is described by a field action $\mathcal{S}_{4D} = \int dt L_{4D}$ with Lagrangian realized as the space integral $\int dxdydz \mathcal{L}_{4D}$. Generally, the Lagrangian density $\mathcal{L}_{4D}$ is a function of the soliton $\boldsymbol{n}(t,x,y,z)$ which is a real 4-component field $[\boldsymbol{n} = (n_1, n_2, n_3, n_4)]$ constrained like $f[\boldsymbol{n}(\xi)] = 1$. For self ineracting field, the typical field expression of $\mathcal{L}_{4D}$ is given by $-\frac{1}{2}(\partial_\mu \boldsymbol{n})^2 - V(\boldsymbol{n}) - \Lambda[f(\boldsymbol{n}) - 1]$ where $V(\boldsymbol{n})$ is a scalar potential; and where the auxiliary field $\Lambda(\xi)$ is a Lagrange multiplier carrying the field constraint. This density $\mathcal{L}_{4D}$ reads in terms of the Hamiltonian as $\Pi.\frac{\partial \mathbf{n}}{\partial t} - \mathcal{H}(\mathbf{n})$. Below, we consider a 4-component skyrmionic field constrained as $\boldsymbol{n}.\boldsymbol{n} = 1$; and focuss on a simple Lagrangian density $\mathcal{L}_\circ = -\frac{1}{2}(\partial_\mu \boldsymbol{n})(\partial^\mu \boldsymbol{n}) - \Lambda[\boldsymbol{n}.\boldsymbol{n} - 1]$ to describe the degrees of freedom of $\boldsymbol{n}$. Being a unit 4-component vector, we can solve the constraint $\boldsymbol{n}.\boldsymbol{n} = 1$ in terms of three angular angles (α, β, γ); by setting

$$\boldsymbol{n} = (\mathbf{m} \sin\gamma,\ \cos\gamma), \qquad \mathbf{m} = (\sin\beta\cos\alpha,\ \sin\beta\sin\alpha,\ \cos\beta) \tag{26}$$

where $\mathbf{m}$ is a unit 3-vector parameterising the unit sphere $\mathbb{S}^2_{[\alpha]}$. Putting this field realisation back into $\mathcal{L}_\circ$, we obtain $\frac{-\cos 2\gamma}{2}(\partial_\mu \gamma)^2 - \frac{1-\cos 2\gamma}{4}(\partial_\mu \mathbf{m})^2 - \Lambda[\mathbf{m}.\mathbf{m} - 1]$. Notice that by restricting the 4D spacetime $\mathbb{R}^{1,3}$ to the 3D hyperplane z=const; and by fixing the component field γ to $\frac{\pi}{2}$, the above Lagrangian density reduces to the one describing the spacetime dynamics of the 2d skyrmion. Notice also that we can expand the differential $d\boldsymbol{n}_A$ in terms of $d\gamma, d\beta, d\alpha$; we find the following

$$d\boldsymbol{n}_a = m_a \cos\gamma d\gamma + \sin\gamma(u_a d\beta + v_a \sin\beta d\alpha), \quad d\boldsymbol{n}_4 = -\sin\gamma d\gamma \tag{27}$$

For convenience, we sometimes refer to the three (α, β, γ) collectively like $\alpha_a = (\alpha_1, \alpha_2, \alpha_3)$; so we have $d\boldsymbol{n}^A = E^A_a d\alpha^b$ with $E^A_a = \frac{\partial \mathbf{n}^A}{\partial \alpha^a}$.

4.2 Conserved topological current

First, we investigate the topological properties of the 3d skyrmion from the target space view; that is without using the spacetine variables $(t,x,y,z) = \xi^\mu$. Then, we turn to study the induced topological properties of the 3d skyrmion viewed from the side of the 4D space time $\mathbb{R}^{1,3}$.

4.2.1 Topological current in target space

The 3d skyrmion field is described by a real four component vector $\boldsymbol{n}_A$ subject to the constraint relation $\boldsymbol{n}_A \boldsymbol{n}^A = 1$; so the soliton has SO(4) $\sim SO(3)_1 \times SO(3)_2$ symmetry leaving invariant the condition $\boldsymbol{n}_A \boldsymbol{n}^A = 1$ that reads explicitly as $(n_1)^2 + (n_2)^2 + (n_3)^2 + (n_4)^2 = 1$. The algebraic condition $f[\boldsymbol{n}] = 1$ induces in turns the constraint equation $df = 0$ leading to $\mathbf{n}^A d\boldsymbol{n}_A = 0$ and showing that $\boldsymbol{n}_A$ and $d\boldsymbol{n}_A$ orthogonal 4-vectors in $\mathbb{R}^4_{(\boldsymbol{n})}$. From this constraint, we can construct $(\boldsymbol{n}^A d\boldsymbol{n}^B - \boldsymbol{n}^B d\boldsymbol{n}^A)/2$ which is a 4×4 antisymmetric matrix $\Omega^{[AB]}$ generating the SO(4) rotations; this $\Omega^{[AB]}$ contains 3+3 degrees of freedom generating the two $SO(3)_1$ and $SO(3)_2$ making SO(4); the first three degrres are given by $\Omega^{[ab]}$ with $a, b = 1, 2, 3$; and the other three concern $\Omega^{[a4]}$. Notice also that, from the view of the target space, the algebraic relation $\boldsymbol{n}_A \boldsymbol{n}^A = 1$ describes a unit 3-sphere $\mathbb{S}^3_{\boldsymbol{n}}$ sitting in $\mathbb{R}^4_{\boldsymbol{n}}$; as such its volume 4-form $d^4\boldsymbol{n}$, which reads as $\frac{1}{4!}\varepsilon_{ABCD}d\boldsymbol{n}^A \wedge d\boldsymbol{n}^B \wedge d\boldsymbol{n}^C \wedge d\boldsymbol{n}^D$, vanishes identically when restricted to the 3-sphere; i.e.: $d^4\boldsymbol{n}|_{\mathbb{S}^3_{\boldsymbol{n}}} = 0$. This vanishing property of $d^4\boldsymbol{n}$ on $\mathbb{S}^3_{\boldsymbol{n}}$ is a key ingredient in the derivation of the topological current $\boldsymbol{J}$ of the 3D skyrmion and its conservation $d\boldsymbol{J} = 0$. Indeed, because of the property $d^2 = 0$ (where we have hidden the wedge product $\wedge$), it follows that $d^4\boldsymbol{n}$ can be expressed as $d\boldsymbol{J}$ with the 3-form $\boldsymbol{J}$ given by

$$\boldsymbol{J} = \frac{1}{4!}\varepsilon_{ABCD}\boldsymbol{n}^A d\boldsymbol{n}^B d\boldsymbol{n}^C d\boldsymbol{n}^D \tag{28}$$

This 3-form describes precisely the topological current in the target space; this is because on $\mathbb{S}^3_{\boldsymbol{n}}$, the 4-form $d^4\boldsymbol{n}$ vanishes; and then $d\boldsymbol{J}$ vanishes. By solving, the skyrmion field constraint $\boldsymbol{n}_A\boldsymbol{n}^A = 1$ in terms of three angles α_a as given by Eq. (26); with these angular coordinates, we have mapping $\boldsymbol{f} : \mathbb{R}^4_{\boldsymbol{n}} \to \mathbb{S}^3_{\boldsymbol{n}}$ with $\mathbb{S}^3_{\boldsymbol{n}} \simeq \mathbb{S}^3_{\alpha}$. By expanding the differentials like $d\boldsymbol{n}^A = E^A_a d\alpha^a$ with $E^A_a = \frac{\partial \boldsymbol{n}^A}{\partial \alpha^a}$; then the conserved current on the 3-sphere $\mathbb{S}^3_\alpha$ reads as follows

$$\boldsymbol{J} = \frac{1}{4!3!}\varepsilon_{ABCD}\left(\boldsymbol{n}^A E^B_b E^C_c E^D_d\right)\varepsilon^{abc}d^3\alpha \tag{29}$$

where we have substituted the 3-form $d\alpha^b d\alpha^c d\alpha^d$ on the 3-sphere $\mathbb{S}^3_\alpha$ by the volume 3-form $\varepsilon^{abc}d^3\alpha$. In this regards, recall that the volume of the 3-sphere is $\int_{\mathbb{S}^3_{(\boldsymbol{n})}} d^3\alpha = \frac{\pi^2}{2}$.

4.2.2 Topological symmetry in spacetime

In the spacetime $\mathbb{R}^{1,3}$ with coordinates $\xi^\mu = (t, x, y, z)$, the 3d skyrmion is described by a four component field $\boldsymbol{n}_A(\xi)$ and is subject to the local constraint relation $\boldsymbol{n}_A\boldsymbol{n}^A = 1$. A typical static configuration of the 3d skyrmion is obtained by solving the field sonctraint in terms of the space coordinates; it is given by Eq. (26) with the local space time fields $\mathbf{m}(\xi)$ and $\gamma(\xi)$ thought of as follows

$$\mathbf{m}(\xi) = \left(\frac{x}{r}, \frac{y}{r}, \frac{z}{r}\right), \quad \gamma(\xi) = \arcsin\frac{2rR}{r^2 + R^2} = \arccos\frac{r^2 - R^2}{r^2 + R^2} \tag{30}$$

with $r = \sqrt{x^2 + y^2 + z^2}$, giving the radius of $\mathbb{S}^2_\xi$, and R associated with the circle $\mathbb{S}^1_\xi$ fibered over $\mathbb{S}^2_\xi$; the value $R = r$ corresponds to $\gamma = \frac{\pi}{2}$ and $R >> r$ to $\gamma = \pi$. Notice that $\gamma(\xi)$ in Eq. (30) has a spherical symmetry as it is a function only of r (no angles α, β, γ). Moreover, as this configuration obeys $\sin\gamma(0) = 0$ and $\sin\gamma(\infty) = 0$; we assume $\gamma(0) = n_0\pi$ and $\gamma(\infty) = n_\infty\pi$. Putting these relations back into (26), we obtain the following configuration

$$\tilde{\boldsymbol{n}} = \left(\frac{2xR}{r^2 + R^2}, \frac{2yR}{r^2 + R^2}, \frac{2zR}{r^2 + R^2}, \frac{r^2 - R^2}{r^2 + R^2}\right) \tag{31}$$

describing a compactification of the space $\mathbb{R}^3_\xi$ into $\mathbb{S}^3_\xi$ which is homotopic to $\mathbb{S}^3_{\boldsymbol{n}}$. From this view, the $\tilde{\boldsymbol{n}} : \xi \to \tilde{\boldsymbol{n}}(\xi)$ is then a mapping from $\mathbb{S}^3_\xi$ into $\mathbb{S}^3_{\mathbf{n}}$ with topological charge given by the winding number characterising the wrapping $\mathbb{S}^3_{\boldsymbol{n}}$ on $\mathbb{S}^3_\xi$; and for which we have the property $\pi_3\left(\mathbb{S}^3_n\right) = \mathbb{Z}$. In this regards, recall that 3-spheres $\mathbb{S}^3$

have a Hopf fibration given by a circle $\mathbb{S}^1$ sitting over $\mathbb{S}^2$ (for short $\mathbb{S}^3 \sim \mathbb{S}^1 \ltimes \mathbb{S}^2$); this non trivial fibration can be viewed from the relation $\mathbb{S}^3 \sim SU(2)$ and the factorisation $U(1) \times SU(2)/U(1)$ with the coset $SU(2)/U(1)$ identified with $\mathbb{S}^2$; and $U(1)$ with $\mathbb{S}^1$. Applying this fibration to $\mathbb{S}^3_{\boldsymbol{\xi}}$ and $\mathbb{S}^3_{\boldsymbol{n}}$, it follows that $\tilde{\boldsymbol{n}} : \mathbb{S}^3_{\xi} \to \mathbb{S}^3_{\boldsymbol{n}}$; and the same thing for the bases $\mathbb{S}^2_{\xi} \to \mathbb{S}^2_{\boldsymbol{n}}$ and for the fibers $\mathbb{S}^1_{\xi} \to \mathbb{S}^1_{\boldsymbol{n}}$. Returning to the topological current and the conserved topological charge $Q = \int_{\mathbb{R}^3} d^3\mathbf{r} J^0(t, \mathbf{r})$, notice that in space time the differential $d\boldsymbol{n}^A$ expands like $(\partial_\mu \boldsymbol{n}^A) d\xi^\mu$; then using the duality relation $J_{[\nu\rho\tau]} = \varepsilon_{\mu\nu\rho\tau} J^\mu$, we find, up to a normalisation by the volume of the 3-sphere $\pi^2/2$, the expression of the topological current $J^\mu(\xi)$ in terms of the 3D skyrmion field

$$J^\mu = \frac{1}{12\pi^2} \varepsilon^{\mu\nu\rho\tau} n^a \partial_\nu n^b \partial_\rho n^c \partial_\tau n^c \varepsilon_{abcd} \tag{32}$$

In terms of the angular variables, this current reads like $\mathcal{N} \partial_\nu \alpha \partial_\rho \beta \partial_\tau \gamma \varepsilon^{\mu\nu\rho\tau}$ with $\mathcal{N} = \frac{1}{2\pi^2} (\sin\beta)(\sin\gamma)^2$. From this current expression, we can determine the associated topological charge Q by space integration over the charge density

$$J^0(t, \mathbf{r}) = -\frac{\sin^2\gamma}{2\pi^2 r^2} \frac{d\gamma}{dr} \tag{33}$$

Because of its spherical symmetry, the space volume $d^3\mathbf{r}$ can be substituted by $4\pi r^2 dr$; then the charge Q reads as the integral $-\frac{4\pi}{2\pi^2} \int_{\gamma(0)}^{\gamma(\infty)} \sin^2\gamma d\gamma$ whose integration leads to the sum of two terms coming from the integration of $\sin^2\gamma = \frac{1}{2} - \frac{1}{2}\cos 2\gamma$. The integral first reads as $\frac{1}{\pi}[\gamma(0) - \gamma(\infty)]$; by substituting $\gamma(0) = n_0\pi$, it contributes like $N\pi$. The integral of the second tem gives $\frac{1}{2\pi}[\sin 2\gamma(0) - \sin 2\gamma(\infty)]$; it vanishes identically. So the topological charge is given by

$$Q = \frac{\gamma(0) - \gamma(\infty)}{\pi} = N \tag{34}$$

5. Effective dynamics of skyrmions

In this section, we investigate the effective dynamics of a point-like skyrmion in a ferromagnetic background field while focussing on the 2d configuration. First, we derive the effective equation of a rigid skyrmion and comment on the underlying effective Lagrangian. We also describe the similarity with the dynamics of an electron in a background electromagnetic field. Then, we study the effect of dissipation on the skyrmion dynamics.

5.1 Equation of a rigid skyrmion

To get the effective equation of motion of a rigid skyrmion, we start by the spin $(0+1)D$ action $\mathcal{S}_{spin} = \int dt L_{spin}$ describing the time evolution of a coherent spin vector modeled by a rotating magnetic moment $\mathbf{n}(t)$ with velocity $\dot{\mathbf{n}} = \frac{d\mathbf{n}}{dt}$; and make some accommodations. For that, recall that the Lagrangian L_{spin} has the structure $L_B - \hbar S \gamma H$ where L_B is the Berry term having the form $L_B = q_e \mathbf{A}.\dot{\mathbf{n}}$ with geometric (Berry) potential $\mathbf{A} \sim \langle \mathbf{n} | \dot{\mathbf{n}} \rangle$; and where H is the Hamiltonian of the magnetic moment $\mathbf{n}(t)$ obeying the constraint $\mathbf{n}^2 = 1$. This magnetisation constraint is solved by two free angles $\alpha(t), \beta(t)$; they appear in the Berry term $L_B = -\hbar S(1 - \cos\beta)\frac{d\alpha}{dt}$. Below, we think of the above magnetisation as a ferromagnetic background $\mathbf{n}(\mathbf{r})$ filling the spatial region of $\mathbb{R}_\xi^{1,d}$ with coordinates $\xi = (t, \mathbf{r})$; and of the skymion as a massive point- like particle $\mathbf{R}(t)$ moving in this background.

5.1.1 Rigid skyrmion

We begin by introducing the variables describing the skyrmion in the magnetic background field $\mathbf{n}(\mathbf{r})$. We denote by M_s the mass of the skyrmion, and by $\mathbf{R}$ and $\dot{\mathbf{R}}$ its space position and its velocity. For concreteness, we restrict the investigation to the spacetime $\mathbb{R}_\xi^{1,2}$ and refer to $\mathbf{R}$ by the components $X_i = (X, Y)$ and to $\mathbf{r}$ by the components $x_i = (x, y)$. Because of the Euclidean metric δ_{ij}; we often we use both notations X^i and $X_i = \delta_{ij}X^j$ without referring to δ_{ij}. Furthermore; we limit the discussion to the interesting case where the only source of displacements in $\mathbb{R}_\xi^{1,2}$ is due to the skyrmion $\mathbf{R}(t)$ (rigid skyrmion). In this picture, the description of the skyrmion $\mathbf{R}(t)$ dissolved in the background magnet $\mathbf{n}(\mathbf{r})$ is given by

$$\mathbf{n}(\mathbf{r}, t) = \mathbf{n}[\mathbf{r} - \mathbf{R}(t)] \tag{35}$$

In this representation, the velocity $\dot{\mathbf{n}}$ of the skyrmion dissolved in the back-ground magnet can be expressed into manners; either like $-\dot{X}^i\frac{\partial \mathbf{n}}{\partial X^i}$; or as $\dot{X}^i\frac{\partial \mathbf{n}}{\partial x^i}$; this is because $\frac{\partial}{\partial X^i} = -\frac{\partial}{\partial x^i}$. With this parametrisation, the dynamics of the skyrmion is described by an action $\mathcal{S}_s = \int dt L_s$ with Lagrangian given by a space integral $L_s = \frac{\hbar s}{\mathrm{a}^2}\int d^2\mathbf{r}\mathcal{L}_s$ and spacetime density as follows

$$\mathcal{L}_s = \gamma\hbar S\mathcal{H} - \hbar S\mathcal{L}_B \tag{36}$$

In this relation, the density $\mathcal{L}_B = -(1 - \cos\beta)\frac{\partial\alpha}{\partial t}$ where now the angular variables are spacetime fields $\beta(t, \mathbf{r})$ and $\alpha(t, \mathbf{r})$. Similarly, the density $\mathcal{H}$ is the Hamiltonian density with arguments as $\mathcal{H}[\mathbf{n}, \partial_\mu\mathbf{n}, \mathbf{r}]$ and magnetic $\mathbf{n}$ as in Eq. (35). In this field action $\mathcal{S}_s$, the prefactor a^{-2} is required by the continuum limit of lattice Hamiltonians H_{latt} living on a square lattice with spacing parameter a. Recall that for these H_{latt}'s, one generally has discrete sums like $\sum_\mu(\dots)$, $\sum_{\mu,\nu}(\dots)$ and so on; in the limit where a is too small, these sums turn into 2D space integrals like $\mathrm{a}^{-2}\int d^2\mathbf{r}(\dots)$. To fix ideas, we illustrate this limit on the typical hamiltonian H_{HDMZ}, it describes the Heisenberg model on the lattice $\mathbb{Z}^2$ augmented by the Dzyaloshinskii-Moriya and the Zeeman interactions [66, 67]

$$H_{HDMZ} = -J\sum_{\langle\mu,\nu\rangle}\mathbf{n}(\mathbf{r}_\mu)\mathbf{n}(\mathbf{r}_\nu) - D\sum_{\mu,\nu,\rho}\mathbf{d}_\mu.[\mathbf{n}(\mathbf{r}_\nu)\wedge\mathbf{n}(\mathbf{r}_\rho)]\varepsilon^{\mu\nu\rho} - \sum_\mu\mathbf{B}.\mathbf{n}(\mathbf{r}_\mu) \tag{37}$$

with $\mathbf{r}_\nu = \mathbf{r}_\mu + \mathrm{a}\mathbf{e}_{\nu\mu}$; that is $\mathbf{e}_{\nu\mu} = (\mathbf{r}_\nu - \mathbf{r}_\mu)/\mathrm{a}$ where a is the square lattice parameter. So, the continuum limit $\mathcal{H}$ of this lattice Hamiltonian involves the target space metric δ_{ab} and the topological Levi-Civita tensor ε_{abc} of the target space $\mathbb{R}_\mathbf{n}^3$; it involves as well the metric $g_{\mu\nu}$ and the completely antisymmetry $\varepsilon_{\mu\nu\rho}$ of the space time $\mathbb{R}_\xi^{1,2}$. In terms of δ_{ab} and ε_{abc} tensors, the continuous hamiltonian density reads as follows

$$\mathcal{H} = \frac{J\mathrm{a}^2}{2}\delta_{ab}\partial^i n^a\partial_i n^b + \mathrm{a}\varepsilon_{abc}d_\mu^a(n^b D_{\nu\rho}n^c)\varepsilon^{\mu\nu\rho} - \mathbf{B}.\mathbf{n} \tag{38}$$

with $\nabla_{\nu\rho} = \mathbf{e}_{\nu\mu}.\nabla$. Below, we set J=1 and, to factorise out the normalisation factor a^2, scale the parameters of the model like $d_\mu^a = a\tilde{d}_\mu^a$ and $\mathbf{B} = \mathrm{a}^2\tilde{\mathbf{B}}$. For simplicity, we sometimes set as well a= 1.

5.1.2 Skyrmion equation without dissipation

To get the effective field equation of motion of the point-like skyrmion without dissipation, we calculate the vanishing condition of the functional variation of the action; that is $\delta(\int dtd^2\mathbf{r}\mathcal{L}) = 0$. General arguments indicate that the effective equation of the skyrmion with topological charge q_s in the background magnet has the form

$$M_s\ddot{X}_i = q_s\mathcal{E}_i + q_s\varepsilon_{ijz}\dot{X}^j\mathcal{B}^z \tag{39}$$

from which one can wonder the effective Lagrangian describing the effective dynamics of the skyrmion. It is given by

$$L_s = \frac{M_s}{2}\delta_{ij}\dot{X}^i\dot{X}^j - q_s\varepsilon_{zij}\mathcal{B}^z X^i\dot{X}^j - q_s\mathcal{V}(X) \tag{40}$$

Notice that the right hand of Eq. (39) looks like the usual Lorentz force $(q_e\mathbf{E} + q_e\dot{\mathbf{r}}\wedge\mathbf{B})$ of a moving electron with q_e in an external electromagnetic field (E_i, B_i); the corresponding Lagrangian is $\frac{m}{2}\dot{\mathbf{r}}^2 + q_e\mathbf{B}.(\mathbf{r}\wedge\dot{\mathbf{r}}) - q_e\mathbf{E}.\mathbf{r}$. This similarity between the skyrmion and the electron in background fields is because the skyrmion has a topological charge q_s that can be put in correspondence with q_e; and, in the same way, the background field magnet $\mathcal{E}_i, \mathcal{B}_i$ can be also put in correspondence with the electromagnetic field (E_i, B_i). To rigourously derive the spacetime Eqs. (39) and (40), we need to perform some manipulations relying on computing the effective expression of $\mathcal{S}_s = \hbar S\int dt(\int d^2\mathbf{r}\mathcal{L}_s)$ and its time variation $\delta\mathcal{S}_s = 0$. However, as $\mathcal{L}_s$ has two terms like $\gamma\hbar S\mathcal{H} - \hbar S\mathcal{L}_B$, the calculations can be split in two stages; the first stage concerns the block $\gamma\hbar S\int d^2\mathbf{r}\mathcal{H}$ with $\mathcal{H}[\mathbf{n}, \partial_\mu\mathbf{n}, \mathbf{r}]$ which is a function of the magnetic texture (35); that is $\mathbf{n}(\mathbf{r} - \mathbf{R})$. The second stage regards the determination of the integral $\hbar S\int d^2\mathbf{r}\mathcal{L}_B$. The computation of the first term is straightforwardly identified; by performing a space shift $\mathbf{r} \to \mathbf{r} + \mathbf{R}$, the Hamiltonian density becomes $\mathcal{H}[\mathbf{n}, \partial_\mu\mathbf{n}, \mathbf{r} + \mathbf{R}]$ with $\mathbf{n}(\mathbf{r})$ and where the dependence in $\mathbf{R}$ becomes explicit; thus allowing to think of the integral $\gamma\hbar S\int d^2\mathbf{r}\mathcal{H}$ as nothing but the scalar energy potential $\mathcal{V}(\mathbf{R}) = \hbar S\gamma\int d^2\mathbf{r}\mathcal{H}(t, \mathbf{r}, \mathbf{R})$. So, we have

$$\frac{\delta}{\delta X^a}\left(\hbar S\gamma\int d^2\mathbf{r}\mathcal{H}\right) = \frac{\partial\mathcal{V}}{\partial X^a} \tag{41}$$

Concerning the calculation of the $\hbar S\int d^2\mathbf{r}\mathcal{L}_B$, the situation is somehow subtle; we do it in two steps; first we calculate the $(\delta\int d^2\mathbf{r}\mathcal{L}_B)$ because we know the variation $\frac{\delta\mathcal{L}_B}{\delta n^a}$ which is equal to $\frac{1}{2}\varepsilon_{abc}n^b\dot{n}^c$. Then, we turn backward to determine $\hbar S\int d^2\mathbf{r}\mathcal{L}_B$ by integration. To that purpose, recall also that the Berry term $\mathcal{L}_B$ is given by $-(1 - \cos\beta)\frac{\partial\alpha}{\partial t}$; and its variation $\frac{\delta\mathcal{L}_B}{\delta n^a}\frac{\partial n^a}{\partial X^j}\partial X^j$ is equal to $\frac{1}{2}\varepsilon_{abc}n^b\dot{n}^c$. To determine the time variation $\delta L_B = \delta\int d^2\mathbf{r}\delta\mathcal{L}_B$, we first expand it like $\int d^2\mathbf{r}\frac{\delta\mathcal{L}}{\delta n^a}\delta n^a$; and use $\delta n^a = -\frac{\partial n^a}{\partial X^j}\partial X^j$ to put it into the form -$\int d^2\mathbf{r}\frac{\delta\mathcal{L}_B}{\delta n^a}\frac{\partial n^a}{\partial X^j}\partial X^j$. Then, substituting $\frac{\delta\mathcal{L}_B}{\delta n^a}$ by its expression $\frac{1}{2}\varepsilon_{abc}n^b\dot{n}^c$ with $\dot{n}^c$ expanded like -$\frac{\partial n^c}{\partial X^i}\dot{X}^i$, we end up with

$$\delta L_B = 2\hbar S\left(\int d^2\mathbf{r}\frac{1}{2}\varepsilon_{abc}n^b\frac{\partial n^c}{\partial x^i}\frac{\partial n^a}{\partial x^j}\right)\varepsilon^{ij}\left(\dot{X}\delta Y - \dot{Y}\delta X\right) \tag{42}$$

Next, using the relation $\varepsilon^{ij}d^2\mathbf{r} = dx^i \wedge dx^j$, the first factor becomes $\int \frac{1}{2}\varepsilon_{abc}n^a dn^b \wedge dn^c$ gives precisely the skyrmion topological charge q_s. So, the resulting δL_B reduces to $2q_s\hbar S\left(\dot{X}\delta Y - \dot{Y}\delta X\right)$ that reads also like

$$\delta L_B = -2q_s\hbar S\varepsilon_{ij}\dot{X}^i\delta X^j \tag{43}$$

This variation is very remarkable because it is contained in the variation of the effective coupling $L_B^{int} = -2q_s\hbar S\varepsilon_{ij}\dot{X}^i X^j$ which can be presented like $L_B^{int} = -q_s\mathcal{A}_i\dot{X}^i$ where we have set $\mathcal{A}_i = 2\hbar S\varepsilon_{zij}X^j$; this vector can be interpreted as the vector potential of an effective magnetic field $\mathcal{B}^z = \frac{1}{2}\varepsilon^{zij}\partial_i\mathcal{A}_i$. By adding the kinetic term $\frac{M_s}{2}\dot{X}^i\dot{X}_i$, we end up with an effective Lagrangian L_B associated with the Berry term; it reads as follows $L_B = \frac{M_s}{2}\delta_{ij}\dot{X}^i\dot{X}^j - q_s\mathcal{A}_i\dot{X}^i$. So, the effective Lagrangian L_{eff} describing the rigid 2d skyrmion in a ferromagnet is

$$L_{eff} = \frac{M_s}{2}\dot{\mathbf{R}}^2 - q_s\mathbf{A}.\dot{\mathbf{R}} - \mathcal{V}(R) \tag{44}$$

From this Lagrangian, we learn the equation of the motion of the rigid skyrmion namely $M_s\ddot{X}_j = f_j + 4q_s\hbar S\varepsilon_{zij}\dot{X}^i$; for the limit $M_s = 0$, it reduces to $\dot{X}^i = \frac{1}{4q_s\hbar S}\varepsilon^{zji}f_j$.

5.2 Implementing dissipation

So far we have considered magnetic moment obeying the constraint $\mathbf{n}^2 = 1$ with time evolution given by the LL equation $\dot{\mathbf{n}} = -\gamma\mathbf{f}\wedge\mathbf{n}$ where the force $\mathbf{f} = -\frac{\partial\mathcal{H}}{\partial\mathbf{n}}$. Using this equation, we deduce the typical properties $\mathbf{n}.\dot{\mathbf{n}} = \mathbf{f}.\dot{\mathbf{n}} = 0$ from which we learn that the time variation $\frac{dH}{dt}$ of the Hamiltonian, which reads as $\hbar S\gamma\int d^2\mathbf{r}\frac{\partial\mathcal{H}}{\partial n^a}\dot{n}^a$, vanishes identically as explicitly exhibited below,

$$\frac{dH}{dt} = -\hbar S\gamma\int d^2\mathbf{r}(\mathbf{f}.\dot{\mathbf{n}}) \tag{45}$$

In presence of dissipation, we loose energy; and so one expects that $\frac{dH}{dt} < 0$; indicating that the rigid skyrmion has a damped dynamics. In what follows, we study the effect of dissipation in the ferromagnet and derive the damped skyrmion equation.

5.2.1 Landau-Lifshitz-Gilbert equation

Due to dissipation, the force $\mathbf{F}$ acting on the rigid skyrmion $\mathbf{R}(t)$ has two terms, the old conservative $\mathbf{f} = -\frac{\partial\mathcal{H}}{\partial\mathbf{n}}$; and an extra force $\delta\mathbf{f}$ linearly dependent in magnetisation velocity $\dot{\mathbf{n}}$. Due to this extra force $\delta\mathbf{f} = -\frac{\alpha}{\gamma}\dot{\mathbf{n}}$, the LL equation gets modified; its deformed expression is obtained by shifting the old force $\mathbf{f}$ like $\mathbf{F} = \mathbf{f} - \frac{\alpha}{\gamma}\dot{\mathbf{n}}$ with α a positive damping parameter (Gilbert parameter). As such, the previous LL relation gives the so called Landau-Lifschitz-Gilbert (LLG) equation [68, 69]

$$\dot{\mathbf{n}} = -\gamma \mathbf{f} \wedge \mathbf{n} + \alpha \dot{\mathbf{n}} \wedge \mathbf{n} \tag{46}$$

where its both sides have $\dot{\mathbf{n}}$. From this generalised relation, we still have $\mathbf{n}.\dot{\mathbf{n}} = 0$ (ensuring $\mathbf{n}^2 = 1$); however $\mathbf{f}.\dot{\mathbf{n}} \neq \mathbf{0}$ as it is equal to the Gilbert term namely $-\alpha(\mathbf{f} \wedge \mathbf{n}).\dot{\mathbf{n}}$. Notice that Eq. (46) still describe a rotating magnetic moment in the target space ($d\mathbf{n} = 0$); but with a different angular velocity Ω which, in addition to $\mathbf{f}$, depends moreover on the Gilbert parameter and the magnetisation $\mathbf{n}$. By factorising Eq. (46) like $\dot{\mathbf{n}} = \Omega \wedge \mathbf{n}$, we find

$$\Omega = \frac{-\gamma}{1+\alpha^2} [\mathbf{f} + \alpha(\mathbf{f} \wedge \mathbf{n})] \tag{47}$$

Notice that in presence of dissipation ($\alpha \neq 0$), the variation of the hamiltonian $\frac{dH}{dt}$ given by (45) is no longer non vanishing; by first replacing $\mathbf{f}.\dot{\mathbf{n}} = -\alpha(\mathbf{f} \wedge \mathbf{n}).\dot{\mathbf{n}}$ and putting back in it, we get

$$\frac{dH}{dt} = \alpha \hbar S \int d^2\mathbf{r} \gamma (\mathbf{f} \wedge \mathbf{n}).\dot{\mathbf{n}} \tag{48}$$

then, substituting $\gamma \mathbf{f} \wedge \mathbf{n} = \alpha \dot{\mathbf{n}} \wedge \mathbf{n} - \dot{\mathbf{n}}$; we find that $\frac{dH}{dt}$ is given by $-\alpha \hbar s \int d^2\mathbf{r}.\dot{\mathbf{n}}^2$ indicating that $\frac{dH}{dt} < 0$; and consequently a decreasing energy $H(t)$ (loss of energy) while increasing time.

5.2.2 Damped skyrmion equation

To obtain the damped skyrmion equation due to the Gilbert term, we consider the rigid magnetic moment $\mathbf{n}[\mathbf{r} - \mathbf{R}(t)]$; and compute the expression of the skyrmion velocity $\dot{\mathbf{R}}$ in terms of the conservative force $\mathbf{f}$ and the parameter α. To that purpose we start from Eq. (46) and multiply both equation sides by $\wedge d\mathbf{n}$ while assuming $\mathbf{f}.\mathbf{n} = 0$ (the conservative force transverse to magnetisation), we get $\dot{\mathbf{n}} \wedge d\mathbf{n} = -\gamma(\mathbf{f}.d\mathbf{n})\mathbf{n} + \alpha(d\mathbf{n}.\dot{\mathbf{n}})\mathbf{n}$. Then, multiply scalarly by $\mathbf{n}$, which corresponds to a projection along the magnetisation, we obtain

$$\mathbf{n}.(\dot{\mathbf{n}} \wedge d\mathbf{n}) = -\gamma(\mathbf{f}.d\mathbf{n}) + \alpha(d\mathbf{n}.\dot{\mathbf{n}}) \tag{49}$$

Substituting $d\mathbf{n}$ and $\dot{\mathbf{n}}$ by their expansions $dx^i(\partial_i\mathbf{n})$ and $-\dot{X}^i(\partial_i\mathbf{n})$, then multiplying by $\wedge dx^l$; we end up with a relation involving $dx^j \wedge dx^l$ (which reads as $\varepsilon^{zjl}d^2\mathbf{r}$); so we have

$$\dot{X}^l\left(4\pi J^0 d^2\mathbf{r}\right) = -\gamma\varepsilon^{0lj}\left(\mathbf{f}.\partial_j\mathbf{n}\right)d^2\mathbf{r} + \alpha\varepsilon^{0lj}\dot{X}_j\left(\partial_j\mathbf{n}\right)^2 d^2\mathbf{r} \tag{50}$$

where we have set $J^0 = \frac{1}{2\pi}\varepsilon^{zij}\mathbf{n}.\left[(\partial_i\mathbf{n}) \wedge (\partial_j\mathbf{n})\right]$, defining the magnetization density, and where we have replaced $\left(\partial_i\mathbf{n}.\partial_j\mathbf{n}\right)$ by $\delta_{ij}(\partial_k\mathbf{n})^2$. By integrating over the 2d space while using $\int J^0 d^2\mathbf{r} = 4\pi q_s$ and setting $\eta_j = \frac{1}{4\pi}\int d^2\mathbf{r}\left(\partial_j\mathbf{n}\right)^2 \equiv \eta$, we arrive at the relation

$$4\pi q_s \dot{X}^l = \gamma\varepsilon^{0lj}\int\left(\mathbf{f}.\partial_j\mathbf{n}\right)d^2\mathbf{r} - 4\pi\eta\alpha\varepsilon^{0lj}\dot{X}^l \tag{51}$$

with $\varepsilon^{zxy} = -\varepsilon_{zxy} = -1$. The remaining step is to replace the conservative force $\mathbf{f}$ by $-\frac{\partial H}{\partial \mathbf{n}}$ and proceeds in performing the integral over $\left(\mathbf{f}.\partial_j\mathbf{n}\right)$. Because of the explicit dependence into $\mathbf{r}$, the $\mathbf{f}.\partial_j\mathbf{n}$ can be expressed like $\partial_j^{\exp}\mathcal{H} - \partial_j^{tot}\mathcal{H}$; the explicit derivation term $\partial_j^{\exp}\mathcal{H}$ has been added because the Hamiltonian density has an explicit dependence $\mathcal{H}[\mathbf{n}, \partial_\mu\mathbf{n}, \mathbf{r}]$. Recall that $\partial_j^{tot}\mathcal{H}$ is given by $\partial_j^{\exp}\mathcal{H} + \frac{\partial\mathcal{H}}{\partial\mathbf{n}}.\partial_j\mathbf{n}$ which is equal to $\partial_j^{\exp}\mathcal{H} - \mathbf{f}.\partial_j\mathbf{n}$. Notice also that the term $\partial_j^{\exp}\mathcal{H}$ can be also expressed like $-\frac{\partial\mathcal{H}}{\partial\mathbf{R}}$. Therefore, the integral $(\mathbf{f}.\partial_j\mathbf{n})d^2\mathbf{r}$ has two contributions namely the $\int\left(\partial_j^{tot}\mathcal{H}\right)d^2\mathbf{r}$ which, being a total derivative, vanishes identically; and the term $\int\left(\partial_j^{\exp}\mathcal{H}\right)d^2\mathbf{r}$ that gives $-\frac{\partial\mathcal{V}}{\partial\mathbf{R}}$. Putting this value back into (51), we end up with

$$4\pi q_s\dot{X}^l = \gamma\varepsilon^{zlj}\left(\frac{\partial\mathcal{V}}{\partial X^j} + \frac{4\pi\eta\alpha}{\gamma}\dot{X}^l\right) \tag{52}$$

Implementing the kinetic term of the skyrmion, we obtain the equation with dissipation $M_s\ddot{\mathbf{R}} = -\frac{\partial\mathcal{V}}{\partial\mathbf{R}} + G\left(\mathbf{z}\wedge\dot{\mathbf{R}} - \frac{\eta\alpha}{q_s}\dot{\mathbf{R}}\right)$ where the constant $G = \frac{4\pi\hbar Sq_s}{a^2}$ stands for the gyrostropic constant.

6. Electron-skyrmion interaction

In this section, we investigate the interacting dynamics between electrons and skyrmions with spin transfer torque (STT) [70]. The electron-skyrmion interaction is given by Hund coupling $J_H(\Psi^\dagger\sigma_a\Psi).n^a$ which leads to emergent $SU(2)$ gauge potential that mediate the interaction between the spin texture $\mathbf{n}(t,\mathbf{r})$ and the two spin states $(\Psi_\uparrow, \Psi_\downarrow)$ of the electron. We also study other aspects of electron/skyrmion system like the limit of large Hund coupling; and the derivation of the effective equation of motion of rigid skyrmions with STT.

6.1 Hund coupling

We start by recalling that a magnetic atom (like iron, manganese, ...) can be modeled by a localized magnetic moment $\mathbf{n}(t,\mathbf{r})$ and mobile carriers represented by a two spin component field $\Psi(t,\mathbf{r})$; the components of the fields $\mathbf{n}$ and Ψ are respectively given by $n^a(t,\mathbf{r})$ with a=1,2,3; and by $\Psi_\alpha(t,\mathbf{r})$ with $\alpha = \uparrow\downarrow$. Using the electronic vector density $\boldsymbol{j}_{(e)} = \Psi^\dagger\sigma\Psi$, the interaction between localised and itinerant electrons of the magnetic atom are bound by the Hund coupling reading as $H_{e-n} = -J_H\mathbf{n}.\boldsymbol{j}_{(e)}$ with Hund parameter $J_H > 0$ promoting alignment of $\mathbf{n}$ and $\boldsymbol{j}_{(e)}$. So, the dynamics of the interacting electron with the backround $\mathbf{n}$ is given by the Lagrangian density $\mathcal{L}_e = \hbar\Psi^\dagger\frac{i\partial}{\partial t}\Psi - H_{e-n}$ expanding as follows

$$\mathcal{L}_e[\Psi,\mathbf{n}] = \hbar\Psi^\dagger\frac{i\partial}{\partial t}\Psi - \Psi^\dagger\left(\frac{\mathbf{P}^2}{2m} - J_H\sigma.\mathbf{n}\right)\Psi \tag{53}$$

where $\mathbf{P} = \frac{\hbar}{i}\nabla$ and $\sigma.\mathbf{n} = \sigma^x n_x + \sigma^y n_y + \sigma^z n_z$.

6.1.1 *Emergent gauge potential*

Because of the ferromagnetic Hund coupling ($J_H > 0$), the spin observable $\hat{S}_e^z = \frac{\hbar}{2}\sigma^z$ of the conduction electron tends to align with the orientation $\sigma^{\mathbf{n}} = \sigma.\mathbf{n}$ of the magnetisation $\mathbf{n}$ — with angle $\theta = (\mathbf{e}_z \hat{,} \mathbf{n})$—; this alinement is accompanied by a local phase change of the electronic wave function Ψ which becomes $\psi = U\Psi$ where $U(t,\mathbf{r}) = e^{i\Theta(t,\mathbf{r})}$ is a unitary $SU(2)$ transformation mapping σ^z into $\sigma^{\mathbf{n}}$; that is $\sigma^{\mathbf{n}} = U^\dagger \sigma^z U$. For later use, we refer to the new two components of the electronic field like $\psi_{+\mathbf{n}}, \psi_{-\mathbf{n}}$ (for short $\psi_{\dot\alpha}$ with label $\dot\alpha = \pm$) such that the gauge transformation reads as $\psi_{\dot\alpha} = U^{\alpha}_{\dot\alpha}\Psi_\alpha$; that is $\psi_\pm = U_{\pm\downarrow}\Psi_\uparrow + U_{\pm\uparrow}\Psi_\downarrow$. This local rotation of the electronic spin wave induces a non abelian gauge potential with components $\mathcal{A}_\mu = -iU\partial_\mu U^\dagger$ mediating the interaction between the electron and the magnetic texture. Indeed, putting the unitary change into $\mathcal{L}_e[\Psi,\mathbf{n}]$, we end up with an equivalent Lagrangian density; but now with new field variables as follows

$$\mathcal{L}_e[\psi,\mathcal{A}_\mu] = \hbar\psi^\dagger\left(i\partial_0 - A^a_0\sigma_a\right)\psi - \psi^\dagger\left(\frac{(\mathbf{P}+\hbar\mathbf{A}^a\sigma_a)^2}{2m} - J_H\sigma^z\right)\psi \tag{54}$$

Here, the vector potential matrix $\mathcal{A}_\mu$ is valued in the $SU(2)$ Lie algebra generated by the Pauli matrices σ^a; so it can be expanded as $A^x_\mu\sigma^x + A^y_\mu\sigma^y + A^z_\mu\sigma^z$ with components $A^a_\mu = \frac{1}{2}Tr(\sigma^a\mathcal{A}_\mu)$. Notice that in going from the old $\mathcal{L}_e[\Psi,\mathbf{n}]$ to the new $\tilde{\mathcal{L}}_e[\psi,\mathcal{A}_\mu]$, the spin texture $\mathbf{n}$ has disappeared; but not completely as it is manifested by an emergent non abelian gauge potential $\mathcal{A}_\mu$; so everything is as if we have an electron interacting with an external field $\mathcal{A}_\mu$. To get the explicit relation between the gauge potential and the magnetisation, we use the isomorphism $SU(2) \sim \mathbb{S}^3$ and the Hopf fibration $\mathbb{S}^1 \times \mathbb{S}^2$ to write the unitary matrix U as follows

$$U = e^{i\gamma}\begin{pmatrix} \cos\frac{\theta}{2} & e^{-i\varphi}\sin\frac{\theta}{2} \\ e^{+i\varphi}\sin\frac{\theta}{2} & -\cos\frac{\theta}{2}\end{pmatrix} \quad , \quad \mathcal{A}_\mu = \begin{pmatrix} \mathfrak{Z}_\mu & W^-_\mu \\ W^+_\mu & -\mathfrak{Z}_\mu \end{pmatrix} \tag{55}$$

where the factor $e^{i\gamma}$ describes $\mathbb{S}^1$ and where, for later use, we have set $W^\pm_\mu = A^1_\mu \pm iA^2_\mu$ and $\mathfrak{Z}_\mu = A^3_\mu$. So, a specific realisation of the gauge transformation is given by fixing $\gamma = cst$ (say $\gamma = 0$); it corresponds to restricting $\mathbb{S}^3$ down to $\mathbb{S}^2$ and SU(2) reduces down to SU(2)/U(1). In this parametrisation, we can also express the unitary matrix U like $\mathbf{m}.\sigma$ with magnetic vector $\mathbf{m} = \left(\sin\frac{\theta}{2}\cos\varphi,\, \sin\frac{\theta}{2}\sin\varphi,\, \cos\frac{\theta}{2}\right)$ obeying the property $\mathbf{m}^2 = 1$; the same constraint as before. By putting back into $U\sigma.\mathbf{n}U^\dagger$, and using some algebraic relations like $\varepsilon_{abd}\varepsilon_{dce} = \delta_{ac}\delta_{be} - \delta_{bc}\delta_{ae}$, we obtain $[2(\mathbf{m}.\mathbf{n})\mathbf{m} - \mathbf{n}].\sigma$. Then, substituting $\mathbf{n}$ by its expression $(\sin\theta\cos\varphi,\, \sin\theta\sin\varphi,\, \cos\theta)$, we end up with the desired direction σ^z appearing in Eq. (54). On the other hand, by putting $U = \mathbf{m}.\sigma$ back into $-iU\partial_\mu U^\dagger$, we obtain an explicit relation between the gauge potential and the magnetic texture namely $A^a_\mu = \varepsilon^{abc}m_b\partial_\mu m_c$. From this expression, we learn the entries of the potential matrix $\mathcal{A}_\mu$ of Eq. (55); the relation with the texture $\mathbf{n}$ is given in what follows seen that $\mathbf{m}(\theta) = \mathbf{n}(\theta/2)$.

6.1.2 Large Hund coupling limit

We start by noticing that the non abelian gauge potential A^a_μ obtained above can be expressed in a condensed form like $\varepsilon^{abc}m_a\partial_\mu m_b$ (for short $\mathbf{m}\wedge\partial_\mu\mathbf{m}$); so it is normal to $\mathbf{m}$; and then it can be expanded as follows

$$A^a_\mu = \frac{1}{2}\mathrm{e}^a\partial_\mu\theta - \mathrm{f}^a\sin\frac{\theta}{2}\partial_\mu\varphi \tag{56}$$

where we have used the local basis vectors $\mathbf{m}(\theta)$, $\mathbf{e}(\theta)$ and $\mathbf{f}(\theta)$. This is an orthogonal triad which turn out to be intimately related with the triad vectors given by Eq. (5); the relationships read respectively like $\mathbf{n}(\theta/2)$, $\mathbf{u}(\theta/2)$ and $\mathbf{v}(\theta/2)$ involving $\theta/2$ angle instead of θ. Substituting these basis vectors by their angular values, we obtain

$$\begin{pmatrix} A_\mu^1 \\ A_\mu^2 \\ A_\mu^3 \end{pmatrix} = \frac{1}{2}\begin{pmatrix} -\sin\varphi \\ \cos\varphi \\ 0 \end{pmatrix}\partial_\mu\theta - \frac{1}{2}\begin{pmatrix} \sin\theta\cos\varphi \\ \sin\theta\sin\varphi \\ \cos\theta - 1 \end{pmatrix}\partial_\mu\varphi \tag{57}$$

from which we learn that the two first components combine in a complex gauge field $W_\mu^\pm = A_\mu^1 \pm iA_\mu^2$ which is equal to $\frac{i}{2}e^{i\varphi}\mathbf{w}^\pm\partial_\mu\mathbf{n}$ with $\mathbf{w}^\pm = \mathbf{u} \pm i\mathbf{v}$; and the third component A_μ^3 has the remarkable form $\frac{1}{2}(1 - \cos\theta)\partial_\mu\varphi$ whose structure recalls the geometric Berry term (7). Below, we set $A_\mu^3 = \mathfrak{Z}_\mu$ as in Eq. (55); it contains the temporal component $\mathfrak{Z}_0$ and the three spatial ones $\mathfrak{Z}_i$ — denoted in Section 2 respectively as a_0 and a_i—.

In the large Hund coupling ($J_H >> 1$), the spin of the electron is quasi- aligned with the magnetisation $\mathbf{n}$; so the electronic dynamics is mainly described by the chiral wave function $(\psi_+, 0)$ denoted below as $\chi = (\chi, 0)$. Thus, the effective properties of the interaction between the electron and the skyrmion can be obtained by restricting the above relations to the polarised electronic spin wave χ. By setting $\psi_- = 0$ into Eq. (54) and using $\chi^\dagger\sigma^x\chi = \chi^\dagger\sigma^y\chi = 0$ and $\chi^\dagger\sigma^z\chi = \bar{\chi}\chi$ as well as replacing $\left(A_\mu^x\sigma_x\right)^2 + \left(A_\mu^y\sigma_y\right)^2$ by $\frac{1}{4}\left(\partial_\mu\mathbf{n}\right)^2$, the Lagrangian (54) reduces to the polarised $\mathcal{L}_e^{(\mathrm{pol})} = \mathcal{L}_e\left[\chi, \mathbf{n}, \mathcal{Z}_\mu\right]$ given by

$$\mathcal{L}_e^{(\mathrm{pol})} = \hbar\chi^\dagger(i\partial_0 - \mathfrak{Z}_0\sigma^z)\chi - \chi^\dagger\left(\frac{\left(P_i + \hbar\mathfrak{Z}_i^a\sigma_a\right)^2}{2m} + \frac{\hbar^2}{8m}\left(\partial_\mu\mathbf{n}\right)^2 - J_H\sigma^z\right)\chi \tag{58}$$

where $(\mathfrak{Z}_0, \mathfrak{Z}_i)$ define the four components of the emergent abelian gauge prepotential $\mathfrak{Z}_\mu$ associated with the Pauli matrix σ^z; their explicit expressions are given by $\mathfrak{Z}_0 = \frac{1}{2}(1 - \cos\theta)\dot{\varphi}$ and $\mathfrak{Z}_i = \frac{1}{2}(1 - \cos\theta)\partial_i\varphi$; their variation with respect to the magnetic texture are related to the magnetisation field like $\frac{\delta\mathfrak{Z}_\mu}{\delta\mathbf{n}} = \frac{1}{2}\partial_\mu\mathbf{n}\wedge\mathbf{n}$.

6.2 Skyrmion with spin transfer torque

Here, we investigate the full dynamics of the electron/skyrmion system $\{e^-, \mathbf{n}\}$ described by the Lagrangian density $\mathcal{L}_{tot}$ containing the parts $\mathcal{L}_\mathbf{n} + \mathcal{L}_{e-\mathbf{n}}$; the electronic Lagrangian $\mathcal{L}_{e-\mathbf{n}}$ is given by Eq. (54). The Lagrangian $\mathcal{L}_\mathbf{n}$, describing the skyrmion dynamics, is as in eqs (5)–(7) namely $-\hbar S\mathfrak{Z}_0 - \mathcal{H}_\mathbf{n}$ with $\mathfrak{Z}_0 = \frac{1}{2}(1 - \cos\theta)\dot{\varphi}$. By setting $\tilde{\mathcal{H}}_\mathbf{n} = \mathcal{H}_\mathbf{n} + \frac{\hbar^2}{8m}\left(\partial_\mu\mathbf{n}\right)^2\psi^\dagger\psi$, the full Lagrangian density $\mathcal{L}_{tot}$ with can be then presented like $\tilde{\mathcal{L}}\left[\psi, \mathfrak{Z}_\mu\right] - \tilde{H}_\mathbf{n}$ like

$$\tilde{\mathcal{L}}\left[\psi, \mathfrak{Z}_\mu\right] = -\hbar S\mathfrak{Z}_0 + \hbar\psi^\dagger\left(i\frac{\partial}{\partial t} - \mathfrak{Z}_0\sigma^z\right)\psi - \psi^\dagger\left(\frac{\left(P_i + \hbar\mathfrak{Z}_i\sigma^z\right)^2}{2m}\right)\psi \tag{59}$$

with $(P_i + \hbar\mathfrak{Z}_i\sigma^z)^2$ expanding as $P_i^2 + \hbar_i^2\mathfrak{Z}^2 + \hbar(P_i\mathfrak{Z}^i + \mathfrak{Z}^iP_i)\sigma^z$. The equations of motion of ψ and $\mathbf{n}$ are obtained as usual by computing the extremisation of this

Lagrangian density with respect to the corresponding field variables. In general, we have $\delta\mathcal{L}_{tot} = (\delta\mathcal{L}_{tot}/\delta\mathbf{n}).\delta\mathbf{n} + (\delta\mathcal{L}_{tot}/\delta\psi).\delta\psi + hc$ which vanishes for $\delta\mathcal{L}_{tot}/\delta\mathbf{n} = 0$ and $\delta\mathcal{L}_{tot}/\delta\psi^\dagger = 0$.

6.2.1 Modified Landau-Lifshitz equation

Regarding the spin texture **n**, the associated field equation of motion is given by $\delta\mathcal{L}_{tot}/\delta\mathbf{n} = 0$; the contributions to this equation of motion come from the variations $\tilde{\mathcal{L}}$ and $\tilde{\mathcal{H}}_\mathbf{n}$ with respect to $\delta\mathbf{n}$ namely

$$\frac{\delta\tilde{\mathcal{H}}}{\delta\mathbf{n}} - \frac{\delta\tilde{\mathcal{L}}}{\delta\mathfrak{Z}_\mu}\frac{\delta\mathfrak{Z}_\mu}{\delta\mathbf{n}} = 0 \tag{60}$$

The variation $\frac{\delta\mathcal{H}_\mathbf{n}}{\delta\mathbf{n}}$ depends on the structure of the skyrmion Hamiltonian density $\tilde{\mathcal{H}}$; its contribution to the equation of motion can be presented like $\lambda\partial^\mu\partial_\mu\mathbf{n} = \mathbf{F}$ with some factor λ. However, the variation $\frac{\delta\mathcal{L}}{\delta\mathfrak{Z}_\mu}\frac{\delta\mathfrak{Z}_\mu}{\delta\mathbf{n}}$ describes skyrmion-electron interaction; and can be done explicitly into two steps; the first step concerns the calculation of the time like component $\frac{\delta\mathcal{L}}{\delta\mathfrak{Z}_0}\frac{\delta\mathfrak{Z}_0}{\delta\mathbf{n}}$; it gives $-\frac{\hbar}{2}[2S + \psi^\dagger\sigma^z\psi](\dot{\mathbf{n}}\wedge\mathbf{n})$; it is normal to **n** and to velocity $\dot{\mathbf{n}}$ and involves the eletron spin density $\rho_e^z = \psi^\dagger\sigma^z\psi$.

The second step deals with the calculation of the space like component $-\frac{\delta\mathcal{L}}{\delta\mathfrak{Z}_i}\frac{\delta\mathfrak{Z}_i}{\delta\mathbf{n}}$; the factor $\frac{\delta\mathcal{L}}{\delta\mathfrak{Z}_i}$ gives $-\hbar\mathcal{J}^i$ with a 3-component current vector density reading as follows

$$\mathcal{J}_i = \frac{1}{2m}\left(\psi^\dagger\sigma^z P_i\psi - P_i\psi^\dagger\sigma^z\psi\right) + \frac{\hbar}{m}\left(\psi^\dagger\psi\right)\mathfrak{Z}_i \tag{61}$$

This vector two remarkable properties: (1) it is given by the sum of two contributions as it it reads like $\mathcal{J}_i^{(+\mathbf{n})} + \mathcal{J}_i^{(-\mathbf{n})}$ with

$$\begin{aligned}\mathcal{J}_l^{(+\mathbf{n})} &= \frac{\hbar}{m}\left(\overline{\psi}_{+\mathbf{n}}\psi_{+\mathbf{n}}\right)\mathfrak{Z}_l + \frac{1}{2m}\left(\overline{\psi}_{+\mathbf{n}}\frac{\hbar}{i}\partial_l\psi_{+\mathbf{n}} - \frac{\hbar}{i}\partial_l\overline{\psi}_{+\mathbf{n}}\psi_{+\mathbf{n}}\right)\\ \mathcal{J}_l^{(-\mathbf{n})} &= \frac{\hbar}{m}\left(\overline{\psi}_{-\mathbf{n}}\psi_{-\mathbf{n}}\right)\mathfrak{Z}_l - \frac{\hbar}{2im}\left(\overline{\psi}_{-\mathbf{n}}\frac{\hbar}{i}\partial_l\psi_{-\mathbf{n}} - \frac{\hbar}{i}\partial_l\overline{\psi}_{-\mathbf{n}}\psi_{-\mathbf{n}}\right)\end{aligned} \tag{62}$$

These vectors are respectively interpreted as two spin polarised currents; the $\mathcal{J}_i^{(+\mathbf{n})}$ is associated with the $\psi_{+\mathbf{n}}$ wave function as it points in the same direction as **n**; the $\mathcal{J}_i^{(-\mathbf{n})}$ is however associated with $\psi_{-\mathbf{n}}$ pointing in the opposite direction of **n**. (2) Each one of the two $\mathcal{J}^{(+\mathbf{n})}$ and $\mathcal{J}^{(-\mathbf{n})}$ are in turn given by the sum of two contributions as they can be respectively split like $\frac{\hbar}{m}\left(\overline{\psi}_{+\mathbf{n}}\psi_{+\mathbf{n}}\right)\mathfrak{Z} + \mathbf{j}_{\psi_{+\mathbf{n}}}$ and $\frac{\hbar}{m}\left(\overline{\psi}_{-\mathbf{n}}\psi_{-\mathbf{n}}\right)\mathfrak{Z} + \mathbf{j}_{\psi_{-\mathbf{n}}}$ with vector density $\mathbf{j}_\psi$ standing for the usual current vector $\mathbf{j}_\psi = \frac{1}{2m}\overline{\psi}\overleftrightarrow{\mathbf{P}}\psi$. The contribution $\frac{\hbar}{m}\left(\overline{\psi}\psi\right)\mathfrak{Z}$ is proportional to the emergent gauge field $\mathfrak{Z}$; it defines a spin torque transfert to the vector current density $\mathcal{J}_i$.

Regarding the factor $\frac{\delta\mathfrak{Z}_i}{\delta\mathbf{n}}$, it gives $\frac{1}{2}(\partial_i\mathbf{n}\wedge\mathbf{n})$; by substituting, the total contribution of $\frac{\delta\mathcal{L}}{\delta\mathfrak{Z}_i}\frac{\delta\mathfrak{Z}_i}{\delta\mathbf{n}}$ leads to $-\frac{\hbar}{2}\left(\mathcal{J}^i\partial_i\mathbf{n}\right)\wedge\mathbf{n}$ that reads in a condensed form like $-\frac{\hbar}{2}(\mathcal{J}.\nabla\mathbf{n})\wedge\mathbf{n}$. Putting back into Eq. (60), we end up with the following modified LL equation

$$-\frac{\hbar}{2}\left[2S + \psi^\dagger\sigma^z\psi\right](\dot{\mathbf{n}}\wedge\mathbf{n}) + \frac{\hbar}{2}(\mathcal{J}.\nabla\mathbf{n})\wedge\mathbf{n} - \frac{\delta H_\mathbf{n}}{\delta\mathbf{n}} = 0 \tag{63}$$

To compare this equation with the usual LL equation ($\hbar S\dot{\mathbf{n}} = \frac{\delta H_\mathbf{n}}{\delta \mathbf{n}} \wedge \mathbf{n}$) in absence of Hund coupling (which corresponds to putting ψ to zero), we multiply Eq. (63) by $\wedge\,\mathbf{n}$ in order to bring it to a comparable relation with LL equation. By setting $\rho_e^z = \psi^\dagger \sigma^z \psi$, describing the electronic spin density $|\psi_{+\mathbf{n}}|^2 - |\psi_{-\mathbf{n}}|^2$; we find

$$-\hbar\left[\frac{S}{\mathrm{a}^{-d}} + \frac{\rho_e^z}{2}\right]\dot{\mathbf{n}} = \frac{\delta H_\mathbf{n}}{\delta \mathbf{n}} \wedge \mathbf{n} - \hbar[(\mathcal{J}.\nabla)\mathbf{n}] \tag{64}$$

where, due to $\mathbf{n}^2 = 1$, the space gradient $\mathcal{J}.\nabla\mathbf{n}$ is normal to $\mathbf{n}$; and so it can be set as $\Omega^{(e)} \wedge \mathbf{n}$ with $\Omega^{(e)} = \mathcal{J}^i \omega_i^{(e)}$. The above equation is a modified LL equation; it describes the dynamics of the spin texture interacting with electrons through Hund coupling. Notice that for $\psi \to 0$, this equation reduces to $\hbar \frac{S}{\mathrm{a}^{-d}} \dot{\mathbf{n}} = \omega^{(n)} \wedge \mathbf{n}$ showing that the vector $\mathbf{n}$ rotates with $\omega^{(n)} = -\frac{\delta H_\mathbf{n}}{\delta \mathbf{n}}$. By turning on ψ, we have $\dot{\mathbf{n}} \sim (\omega^{(n)} + \Omega^{(e)}) \wedge \mathbf{n}$ indicating that the LL rotation is drifted by $\Omega^{(e)}$ coming from two sources: (*i*) the term $\hbar[(\mathcal{J}.\nabla)\mathbf{n}]$ which deforms LL vector $\omega^{(n)}$ drifted by the $\mathbf{n} \wedge (\mathcal{J}.\nabla\mathbf{n})$; and (*ii*) the electronic spin density $\rho_e^z = \frac{N_e}{\mathrm{a}^{-d}}$; this term adds to the density $\frac{S}{\mathrm{a}^{-d}}$ of the magnetic texture per unit volume; it involves the number $N_e = N_e^{+\mathbf{n}} - N_e^{-\mathbf{n}}$ with $N_e^{\pm\mathbf{n}}$ standing for the filling factor of polarized conduction electrons. Moreover, if assuming $\mathbf{n}(t, \mathbf{r}) = \mathbf{n}(\mathbf{r} - \mathbf{V}_s t)$ with a uniform $\mathbf{V}_s$, then the drift velocity $\dot{n}^a = -(\partial_i n^a) V_s^i$ and $(J_e^i \partial_i) n^a = J_e^a$. Putting back into the modified LLG equation, we end up with the following relation between the $\mathbf{V}_s$ and $\mathbf{v}_e$ velocities $(S + \frac{n_e}{2}) v_s^a = n_e v_e^a$ where we have set $(\partial_i n^a) V_s^i = v_s^a$ and $J_e^a = n_e v_e^a$.

6.2.2 Rigid skyrmion under spin transfer torque

Here, we investigate the dynamics of a 2D rigid skyrmion $[\mathbf{n} = \mathbf{n}(\mathbf{r} - \mathbf{R})]$ under a spin transfer torque (STT) induced by itinerant electrons. For that, we apply the method, used in sub-subSection 5.1.2 to derive L_s from the computation space integral of $\int d^2\mathbf{r}\mathcal{L}_s$ and Eq. (36). To begin, recall that in absence of the STT effect, the Lagrangian L_s of the 2D skyrmion's point- particle, with position $\mathbf{R} = (X, Y)$ and velocity $\dot{\mathbf{R}} = (\dot{X}, \dot{Y})$, is given by $\frac{M_s}{2}\dot{\mathbf{R}}^2 - \frac{G}{2}\mathbf{z}.(\mathbf{R} \wedge \dot{\mathbf{R}}) - V(\mathbf{R})$ with effective scalar energy potential $V(\mathbf{R}) = \int d^2\mathbf{r}\mathcal{H}(\mathbf{r}, \mathbf{R})$ and a constant $G = \frac{4\pi\hbar}{\mathrm{a}^2} q_s S$. Under STT induced by Hund coupling, the Lagrangian L_s gets deformed into $\tilde{\mathrm{L}}_s = L_s + \Delta L_s$, that is

$$\tilde{\mathrm{L}}_s = \frac{M_s}{2}\dot{\mathbf{R}}^2 - \frac{G}{2}\mathbf{z}.(\mathbf{R} \wedge \dot{\mathbf{R}}) - V(\mathbf{R}) + \Delta L_s \tag{65}$$

To determine ΔL_s, we start from $\tilde{\mathrm{L}}_s = \int d^2\mathbf{r}\tilde{\mathcal{L}}_{tot}$ with Lagrangian density as $\tilde{\mathcal{L}}_{tot} = \tilde{\mathcal{L}} - \tilde{\mathcal{H}}_\mathbf{n}$ with $\tilde{\mathcal{L}}$ given by Eq. (59). For convenience, we set $\tilde{\mathcal{L}} = -\hbar S \mathfrak{Z}_0 + \tilde{\mathcal{L}}_{e-\mathbf{n}}$ and set

$$\tilde{\mathcal{L}}_{e-\mathbf{n}} = \hbar\psi^\dagger\left(i\frac{\partial}{\partial t} - \mathfrak{Z}_0\sigma^z\right)\psi - \psi^\dagger\left(\frac{(P_i + \hbar\mathfrak{Z}_i\sigma^z)^2}{2m}\right)\psi \tag{66}$$

The deviation ΔL_s with respect to L_s in (65) comes from those terms in Eq. (66). Notice that this expression involves the wave function ψ coupled to the emergent gauge potential field $\mathfrak{Z}_\mu = (\mathfrak{Z}_0, \mathfrak{Z}_i)$; that is $-\hbar\int d^2\mathbf{r}\psi^\dagger\sigma^z\psi\mathfrak{Z}_0$ and

$-\frac{1}{2m}\int d^2\mathbf{r}\psi^\dagger\left[(P_i+\hbar\mathfrak{Z}_i\sigma^z)^2\right]\psi$. Thus, to obtain ΔL_s, we first calculate the variation $\frac{\delta(\Delta L_s)}{\delta\mathfrak{Z}_\mu}\delta\mathfrak{Z}_\mu$ and put $\delta\mathfrak{Z}_\mu=\frac{\delta\mathfrak{Z}_\mu}{\delta\mathbf{R}}.\delta\mathbf{R}$. Once, we have the explicit expression of this variation, we turn backward to deduce the value of ΔL_s. To that purpose, we proceed in two steps as follows: (*i*) We calculate the temporal contribution $\frac{\delta(\Delta L_s)}{\delta\mathfrak{Z}_0}\frac{\delta\mathfrak{Z}_0}{\delta\mathbf{R}}.\delta\mathbf{R}$; and (*ii*) we compute the spatial $\frac{\delta(\Delta L_s)}{\delta\mathfrak{Z}_i}\frac{\delta\mathfrak{Z}_i}{\delta\mathbf{R}}.\delta\mathbf{R}$. Using the variation $\delta\mathfrak{Z}_0=\frac{1}{2}\delta\mathbf{n}.(\mathbf{n}\wedge\partial_j\mathbf{n})\dot{X}^j$, the contribution of the first term can be put as follows

$$\frac{\delta(\Delta L_s)}{\delta\mathfrak{Z}_0}\frac{\delta\mathfrak{Z}_0}{\delta X^l}\delta X^l=-\frac{\hbar}{2}\mathbf{J}_0^z\varepsilon_{zij}\left[\dot{X}^i\delta X^j\right] \tag{67}$$

where we have set $\rho^z=\psi^\dagger\sigma^z\psi$ and $\mathbf{J}_0^z=\int d^2\mathbf{r}\frac{\rho^z}{2}\varepsilon^{zkl}\mathbf{n}.(\partial_k\mathbf{n}\wedge\partial_l\mathbf{n})$. Notice that the right hand side in above relation can be also put into the form $\frac{\hbar}{2}J_0^z\varepsilon_{zij}\left[\delta\dot{X}^iX^j\right]-\delta\left[\frac{\hbar}{2}\varepsilon_{zij}J_0^z\dot{X}^iX^j\right]$ indicating that ΔL_s must contain the term $\frac{\hbar}{2}\varepsilon_{zij}J_0^z\dot{X}^iX^j$ which reads as well like $\frac{\hbar}{2}J_0\mathbf{z}.(\dot{\mathbf{R}}\wedge\mathbf{R})$. Regarding the spatial part $\frac{\delta(\Delta L_s)}{\delta\mathfrak{Z}_i}.\frac{\delta\mathfrak{Z}_i}{\delta X^l}\delta X^l$, we have quite similar calculations allowing to put it in the following form

$$\frac{\delta(\Delta L_s)}{\delta\mathfrak{Z}_i}.\frac{\delta\mathfrak{Z}_i}{\delta X^l}\delta X^l=-\hbar\varepsilon_{zij}J^{zi}\delta X^j \tag{68}$$

where we have set $J^{zi}(t)=\int d^2\mathbf{r}\mathcal{J}^{zi}(t,\mathbf{r})$ with $\mathcal{J}^{zi}(t,\mathbf{r})$ given by Eq. (61). Here also notice that the right hand of above equation can be put as well like $\delta\left[-\hbar\varepsilon_{zij}J^{zi}X^j\right]$ indicating that ΔL_s contains in addition to $\frac{\hbar}{2}J_0\mathbf{z}.(\dot{\mathbf{R}}\wedge\mathbf{R})$, the term $-\hbar\varepsilon_{zij}J^{zi}X^j$ which reads also as $-\hbar\mathbf{z}.\boldsymbol{J}\wedge\mathbf{R}$ with two component vector $\boldsymbol{J}=(J^{zx},J^{zy})$. Thus, we have the following modified skyrmion equation

$$\tilde{L}_s=\frac{M_s}{2}\dot{\mathbf{R}}^2-\frac{1}{2}(G+\hbar J_0)\mathbf{z}.(\dot{\mathbf{R}}\wedge\mathbf{R})+\hbar\mathbf{z}.(\boldsymbol{J}\wedge\mathbf{R})-V(\mathbf{R}) \tag{69}$$

from which we determine the modified equation of motion of the rigid skyrmion in presence of spin transfer torque.

7. Comments and perspectives

In this bookchapter, we have studied the basic aspects of the solitons dynamics in various $(1+d)$ spacetime dimensions with $d=1,2,3$; while emphasizing the analysis of their topological properties and their interaction with the environment. After having introduced the quantum SU(2) spins, their coherent vector representation $\mathbf{S}=\mathcal{R}(\alpha,\beta,\gamma)\mathbf{S}_0$ with $\mathbf{S}_0$ standing for the highest weight spin state; and their link with the magnetic moments $\mu\propto S\mathbf{n}$, we have revisited the time evolution of coherent spin states; and proceeded by investigating their spatial distribution while focusing on kinks, 2d and 3d skyrmions. We have also considered the rigid skyrmions dissolved in the magnetic texture without and with dissipation. Moreover, we explored the interaction between electrons and skyrmions and analyzed the effect of the spin transfer torque. In this regard, we have refined the results concerning the modified LL equation for the rigid skyrmion in connection with emergent non abelian SU(2) gauge fields. It is found that the magnetic skyrmions, existing in a ferromagnetic (FM) medium, show interesting behaviors such as emergent electrodynamics [71] and current-driven motion at low current densities

[72, 73]. Consequently, the attractive properties of ferromagnetic skyrmions make them promising candidates for high-density and low-power spintronic technology. Besides, ferromagnetic skyrmions have the potential to encode bits in low-power magnetic storage devices. Therefore, alternative technology of forming and controlling skyrmions is necessary for their use in device engineering. This investigation was performed by using the field theory method based on coherent spin states described by a constrained spacetime field captured by $f(\mathbf{n}) = 1$. Such condition supports the topological symmetry of magnetic solitons which is found to be characterised by integral topological charges Q that are interpreted in terms of magnetic skyrmions and antiskyrmion; these topological states can be imagined as (winding) quasiparticle excitations with $Q > 0$ and $Q < 0$ respectively.

Regarding these two skyrmionic configurations, it is interesting to notice that, unlike magnetic skyrmions, the missing rotational symmetry of antiskyrmions leads to anisotropic DMI, which is highly relevant for racetrack applications. It follows that antiskyrmions exist in certain Heusler materials having a particular type of DMI, including MnPtPdSn [36] and MnRhIrSn [74]. It is then deduced that stabilized antiskyrmions can be observed in materials exhibiting D_{2d} symmetry such as layered systems with heavy metal atoms. Furthermore, the antiskyrmion show some interesting features, namely long lifetimes at room temperature and a parallel motion to the applied current [75]. Thus, antiskyrmions are easy to detect using conventional experimental techniques and can be considered as the carriers of information in racetrack devices.

To lift the limitations associated with ferromagnetic skyrmions for low-power spintronic devices, recent trends combine multiple subparticles in different magnetic surroundings. Stable room-temperature antiferromagnetic skyrmions in synthetic Pt/Co/Ru antiferromagnets result from the combination of two FM nano-objects coupled antiferromagnetically [76]. Compared to their ferromagnetic analogs, antiferromagnetic skyrmions exhibit different dynamics and are driven with several kilometers per second by currents. Coupling two subsystems with mutually reversed spins, gives rise to ferrimagnetic skyrmions as detected in GdFeCo films using scanning transmission X-ray microscopy [77]. At ambient temperature, these skyrmions move at a speed of $50\,m/s$ with a reduced skyrmion Hall angle of 20°. Characterized by uncompensated magnetization, the vanishing angular momentum line can be utilized as a self-focusing racetrack for skyrmions. Another technologically promising object is generated by the coexistence of skyrmions and antiskyrmions in materials with D_{2d} symmetry. The resulting spin textures constitute information bits ‘0’ and ‘1’ generalizing the concept of racetrack device. Insensitive to the repulsive interaction between the two distinct nano-objects, such emergent devices are promising solution for racetrack storage applications.

Acknowledgements

L. B. Drissi would like to acknowledge “Académie Hassan II des Sciences et Techniques-Morocco”. She also acknowledges the Alexander von Humboldt Foundation for financial support via the George Forster Research Fellowship for experienced scientists (Ref. 3.4 - MAR - 1202992).

Author details

Lalla Btissam Drissi[1,2*], El Hassan Saidi[1,2], Mosto Bousmina[2,3] and Omar Fassi-Fehri[2]

1 LPHE, Modeling and Simulations, Faculty of Science, Mohammed V University in Rabat, Morocco

2 Hassan II Academy of Science and Technology, Rabat, Morocco

3 Euromed Research Institute, Euro-Mediterranean University of Fes, Fes, Morocco

*Address all correspondence to: lalla-btissam.drissi@um5.ac.ma; b.drissi@academiesciences.ma

References

[1] B. Göbel, I. Mertig, O. A. Tretiakov, Beyond skyrmions: Review and perspectives of alternative magnetic quasiparticles, Physics Reports 895, 1 (2021).

[2] N. Nagaosa, and Y. Tokura, 2013, "Topological properties and dynamics of magnetic Skyrmions", Nature Nanotechnol. 8, 899-911

[3] L. B. Drissi, E. H. Saidi, A signature index for third order topological insulators, J Phys Condens Matter. (2020), 32(36): 365704.

[4] L. B. Drissi, E. H. Saidi, Domain walls in topological tri-hinge matter, Eur. Phys. J. Plus (2021) 136: 68.

[5] J. Sampaio, V. Cros, S. Rohart, A. Thiaville, A. Fert, Nucleation, stability and current-induced motion of isolated magnetic skyrmions in nanostructures, Nature Nanotechnology 8 (2013) 839.

[6] A. Fert, V. Cros, J. Sampaio, Skyrmions on the track, Nat. Nanotechnol. 8 (2013) 152–156.

[7] G. Yang, P. Stano, J. Klinovaja, D. Loss, Majorana bound states in magnetic skyrmions, Physical Review B 93 (2016) 224505.

[8] K. M. Hals, M. Schecter, M. S. Rudner, Composite topological excitations in ferromagnet-superconductor heterostructures, Physical Review Letters 117 (2016) 017001.

[9] G. Yu, P. Upadhyaya, Q. Shao, H. Wu, G. Yin, X. Li, C. He, W. Jiang, X. Han, P. K. Amiri, et al., Room-temperature skyrmion shift device for memory application, Nano Letters 17 (2017) 261–268.

[10] I. L. Fernandes, J. Bouaziz, S. Blügel, S. Lounis, Universality of defect-skyrmion interaction profiles, Nature Communications 9 (2018) 4395.

[11] G. Finocchio, F. Buttner, R. Tomasello, M Carpentieri, M. Klaui, (2016) "Magnetic Skyrmions: from fundamental to applications", J. Phys. D. Appl. Phys, 49, 423001.

[12] X. Zhang, Y. Zhou, M. Ezawa, G. P. Zhao, W. Zhao, (2015) "Magnetic Skyrmion transistor: Skyrmion motion in a voltage-gated nanotrack", Sc. Rep. 5, 11369.

[13] T. H. R. Skyrme, "A unified field theory of mesons and baryons", Nucl. Phys. 31, (1962) 556-569.

[14] S. Sondhi, A. Karlhede, S. Kivelson, E. Rezayi, Skyrmions and the crossover from the integer to fractional quantum Hall effect at small Zeeman energies, Physical Review B 47 (1993) 16419.

[15] Y. Ohuchi, et al. 2018, "Electric-field control of anomalous and topological Hall effects in oxide bilayer thin films," Nat. Commun. 9, 213

[16] U. Al Khawaja, H. Stoof, Skyrmions in a ferromagnetic Bose–Einstein condensate, Nature 411 (2001) 918.

[17] J. Fukuda, S. Žumer, Quasi-two-dimensional skyrmion lattices in a chiral nematic liquid crystal, Nature Communications 2 (2011) 246.

[18] S. L. Sondhi, A. Karlhede, S. A. Kivelson, E. H. Rezayi, PRB. B47, 16419 (1993).

[19] S. E. Barrett, G. Dabbagh, L. N. Pfeiffer, K. W. West, R. Tycko, PRL 74, 5112 (1995).

[20] V. F. Mitrovic, M. Horvatic, C. Berthier, S. A. Lyon, M. Shayegan, PRB 76, 115335 (2007).

[21] L. Brey, H. A. Fertig, R. Côté, A. H. Mac Donald, PRL 75, 2562 (1995).

[22] Y. Gallais, J. Yan, A. Pinczuk, L. N. Pfeiffer, K. W. West, PRL 100, 086806 (2008).

[23] H. Zhu, G. Sambandamurthy, Y. P. Chen, P. Jiang, L. W. Engel, D. C. Tsui, L. N. Pfeiffer, K. W. West, PRL 104, 226801 (2010).

[24] A. Fert, N. Reyren, and V. Cros, Nat. Rev. Mater. 2, 1 (2017).

[25] H. Vakili, Y. Xie, and A. W. Ghosh, Phys. Rev. B 102, 174420 (2020).

[26] S. Muhlbauer, B. Binz, F. Joinetz, C. Pfleiderer, A. Rosch, A. Neubauer, *R. georgii*, and P. Boni, Science 323, 915 (2009).

[27] X. Z. Yu, Y. Onose, N. Kanazawa, J. H. Park, J. H. Han, Y. Matsui, N. Nagaosa, and Y. Tokura, Nature (London) 465, 901 (2010).

[28] X. Z. Yu, N. Kanazawa, Y. Onose, K. Kimoto, W. Z. Zhang, S. Ishiwata, Y. Matsui, and Y. Tokura, Nat. Mater. 10, 106 (2010)

[29] T. Tanigaki, K. Shibata, N. Kanazawa, X. Yu, Y. Onose, H. S. Park, D. Shindo, and Y. Tokura, Nano Lett. 15, 5438 (2015).

[30] I. Dzyaloshinsky, A thermodynamic theory of weak ferromagnetism of antiferromagnetics, J. Phys. Chem. Sol. 4 (1958) 241–255.

[31] T. Moriya, Anisotropic superexchange interaction and weak ferromagnetism, Phys. Rev. 120 (1960) 91.

[32] M. V. Mohammad and S. Satpathy, "Dzyaloshinskii-Moriya interaction in the presence of Rashba and Dresselhaus spin-orbit coupling", Phys. Rev. B, 97 (2018) 094419.

[33] N. Nagaosa and Y. Tokura, Nat. Nanotechnol. 8, 899 (2013).

[34] S. Heinze, K. Von Bergmann, M. Menzel, J. Brede, A. Kubetzka, R. Wiesendanger, G. Bihlmayer, S. Blügel, Spontaneous atomic-scale magnetic skyrmion lattice in two dimensions, Nature Physics 7 (2011) 713–718.

[35] J. A. Garlow, S. D. Pollard, M. Beleggia, T. Dutta, H. Yang, Y. Zhu, Physical Review Letters 122 (2019) 237201.

[36] A. K. Nayak, V. Kumar, T. Ma, P. Werner, E. Pippel, R. Sahoo, F. Damay, U. K. Röß ler, C. Felser, S. S. Parkin,, Nature 548 (2017) 561.

[37] A. Leonov, M. Mostovoy, Multiply periodic states and isolated skyrmions in an anisotropic frustrated magnet, Nat. Commun. 6 (2015) 8275.

[38] R. Ozawa, S. Hayami, Y. Motome, Zero-field skyrmions with a high topological number in itinerant magnets, Physical Review Letters 118 (2017) 147205.

[39] C. Lee, J. L. Yang, et al., Monolayer honeycomb structures of group-IV elements and III-V binary compounds: First-principles calculations. Physical Review B. 15 (2010) 155453.

[40] L. B. Drissi, E. H. Saidi, M. Bousmina, O. Fassi-Fehri, Journal of Physics: Condensed Matter, 24(48), (2012) 485502

[41] L. B. Drissi, N. B. Kanga, S. Lounis, F. Djeffal, and S. Haddad, Electron-phonon dynamics in 2D carbon based-hybrids XC (X= Si, Ge, Sn), Journal of Physics: Condensed Matter. 31 (2019) 135702.

[42] W. Yao, D. Xiao, and Q. Niu, Valley-dependent optoelectronics from inversion symmetry breaking, Physical Review B. 77 (2019) 235406.

[43] J. Liu, M. Shi, J. Lu, and M. P. Anantram, Analysis of electrical-field-dependent Dzyaloshinskii-Moriya

interaction and magnetocrystalline anisotropy in a two-dimensional ferromagnetic monolayer. Physical Review B. 97 (2019) 054416.

[44] L. B. Drissi, K. Sadki, M.H. Kourra, and M. Bousmina, Strain-engineering of Janus SiC monolayer functionalized with H and F atoms. Physical Chemistry Chemical Physics. 123 (2018) 185106.

[45] Y. Mao, H. Xu, J. Yuan, and J. Zhong. Functionalization of the electronic and magnetic properties of silicene by halogen atoms unilateral adsorption: a first-principles study. Journal of Physics: Condensed Matter 30 (2018) 365001.

[46] L. B. Drissi, F. Z. Ramadan, and S. Lounis. Halogenation of SiC for band-gap engineering and excitonic Functionalization. Journal of Physics: Condensed Matter. 29 (2017) 455001.

[47] M. Sun, Q. Ren, Y. Zhao, J. P. Chou, J. Yu, and W. Tang, Electronic and magnetic properties of 4d series transition metal substituted graphene: a first-principles study. Carbon 120 (2017) 265-273.

[48] L. B. Drissi, F. Z. Ramadan, and N. B-J Kanga. Fluorination-control of electronic and magnetic properties in GeC-hybrid. Chemical Physics Letters. 659 (2016) 148-153.

[49] E. J. Kan, H. J. Xiang, F. Wu, C. Tian, C. Lee, J. L. Yang, et al. Prediction for room-temperature half-metallic ferromagnetism in the half-fluorinated single layers of BN and ZnO. Applied Physics Letters. 97 (2010) 122503.

[50] Y. Ma, Y. Dai, M. Guo, C. Niu, L. Yu, and B. Huang. Magnetic properties of the semifluorinated and semihydrogenated 2D sheets of group-IV and III-V binary compounds. Applied Surface Science. 257 (2018) 7845-7850.

[51] A. N. Mazurenko, S. A. Rudenko, D. S. Nikolaev, A. Medvedeva, I. Lichtenstein, and M. I. Katsnelson. Role of direct exchange and Dzyaloshinskii-Moriya interactions in magnetic properties of graphene derivatives: C_2F and C_2H. Physical Review B. 94 (2016) 214411.

[52] D. I. Badrtdinov, S. A. Nikolaev, A. N. Rudenko, M. I. Katsnelson, and V. V. Mazurenko. Nanoskyrmion engineering with s p-electron materials: Sn monolayer on a SiC (0001) surface. Physical Review B. 98 (2018) 184425

[53] F. Z. Ramadan, F. J dos Santos, L. B. Drissi and S. Lounis, Complex magnetism of the two-dimensional antiferromagnetic Ge2F: from a Neel spin-texture to a potential antiferromagnetic skyrmion, RCS (2021).

[54] R. Ahl Laamara, L. B. Drissi, E. H. Saidi, D-string fluid in conifold, I: Topological gauge model, Nucl. Phys. B, 743, (2006), 333-353; D-string fluid in conifold: II. Matrix model for D-droplets on S^3, Nucl. Phys. B 749 (2006) 206-224.

[55] E. H. Saidi, Quantum line operators from Lax pairs, Jour of Math Physics 61, 063501 (2020); Gapped gravitinos, isospin 1/2 particles, and N=2 partial breaking, Progress of Theoretical and Experimental Physics, 2019, 013B01, https://doi.org/10.1093/ptep/pty144.

[56] D. Finkelstein, J. Rubinstein, Connection between spin, statistics, and kinks, Journal of Mathematical Physics 9 (1968) 1762–1779.

[57] G. S. Adkins, C. R. Nappi, E. Witten, Static properties of nucleons in the Skyrme model, Nuclear Physics B 228 (1983) 552–566.

[58] J. Xia, X. Zhang, M. Ezawa, Z. Hou, W. Wang, X. Liu, Y. Zhou, Current-driven dynamics of frustrated skyrmions in a synthetic antiferromagnetic bilayer, Physical Review Applied 11 (2019) 044046.

[59] M. Weiß enhofer, U. Nowak, Orientation-dependent current-induced motion of skyrmions with various topologies, Physical Review B 99 (2019) 224430

[60] J. H. Han, Skyrmions in CMP, Spinger Tracts in Modern Physics 278

[61] L. D. Landau, E.M Lifshitz, Theory of the dispersion of magnetic permeability in ferromagnetic bodies. Phys, Z. Sowietunion 8, 153 (1935).

[62] M. V. Berry; Quantal phase factors accompaying adiabatic changes. Proc. R. Soc. Lond. A 392, 45 (1984).

[63] A. Hatcher, (2002), Algebraic Topology, Cambridge University Press, ISBN 978-0-521-79540-1, MR 1867354.

[64] S. Coleman, "The quantum sine-Gordon equation as the massive Thirring model", Phys. Rev. D 11, 2088 (1975).

[65] A. M. Kosevich, B. Ivanov, A. Kovalev, Magnetic solitons, Physics Reports 194 (1990) 117–238.

[66] S. Huang, C. Zhou, G. Chen, H. Shen, A. K. Schmid, K. Liu, Y. Wu, Stabilization and current-induced motion of antiskyrmion in the presence of anisotropic Dzyaloshinskii-Moriya interaction, Phys. Rev. B 96 (2017) 144412.

[67] K.-W. Kim, K.-W. Moon, N. Kerber, J. Nothhelfer, K. Everschor-Sitte, Asymmetric skyrmion Hall effect in systems with a hybrid Dzyaloshinskii-Moriya interaction, Physical Review B 97 (2018) 224427.

[68] T. Gilbert, A Lagrangian formulation of the gyromagnetic equation of the magnetization field, Physical Review 100 (1955) 1243.

[69] Y. Nakatani, Y. Uesake, and N. Hayashi, "Direct solution of the Landau-Lifshitz-Gilbertequation for micromagnetics", Jpn. J. Appl. Phys., Vol. 28, 1989, pp. 2485-2507.

[70] D. C. Ralph, M. D. Stiles, Spin Transfer Torques, J. Magn. Magn. Mater. 320, 1190-1216 (2008).

[71] P. Bruno, V. Dugaev, M. Taillefumier, Topological Hall effect and Berry phase in magnetic nanostructures, Phys. Rev. Lett. 93 (2004) 096806.

[72] T. Dohi, S. DuttaGupta, S. Fukami, H. Ohno, Formation and current-induced motion of synthetic antiferromagnetic skyrmion bubbles, Nature Communications 10 (2019) 5153.

[73] X. Yu, Y. Tokunaga, Y. Kaneko, W. Zhang, K. Kimoto, Y. Matsui, Y. Taguchi, Y. Tokura, Biskyrmion states and their current-driven motion in a layered manganite, Nature Communications 5 (2014) 3198.

[74] J. Jena, R. Stinshoff, R. Saha, A. K. Srivastava, T. Ma, H. Deniz, P. Werner, C. Felser, S. S. Parkin, Observation of magnetic antiskyrmions in the low magnetization ferrimagnet Mn2Rh0.95Ir0.05Sn, Nano Letters 20 (2019) 59.

[75] M. Hoffmann, B. Zimmermann, G. P. Müller, D. Schürhoff, N. S. Kiselev, C. Melcher, S. Blügel, Antiskyrmions stabilized at interfaces by anisotropic Dzyaloshinskii-Moriya interactions, Nat. Commun. 8 (2017) 308.

[76] W. Legrand, D. Maccariello, F. Ajejas, S. Collin, A. Vecchiola, K. Bouzehouane, N. Reyren, V. Cros, A. Fert, Room-temperature stabilization of antiferromagnetic skyrmions in synthetic antiferromagnets, Nature Materials 19 (2020) 34.

[77] S. Woo, K. M. Song, X. Zhang, Y. Zhou, M. Ezawa, X. Liu, S. Finizio, J. Raabe, N. J. Lee, S.-I. Kim, et al., Current-driven dynamics and inhibition of the skyrmion Hall effect of ferrimagnetic skyrmions in GdFeCo films, Nature Communications 9 (2018) 959.

Chapter 2

Skyrmions in Thin Films, Interfaces and Antiferromagnetism

Rajesh Kumar Rajagopal

Abstract

Magnetic skyrmions are small whirling topological defects in a texture magnetization state. Their stabilization and dynamics depend strongly on their topological properties. Skyrmions are induced by non-centrosymmetric crystal structure of magnetic compounds and thin films. Skyrmions are extremely small, with diameters in the nanometer range, and behave as particles that can be created, moved and annihilated. This makes them suitable for information storage and logic technologies. Skyrmions had been observed only at low temperature, and mostly under large applied magnetic fields. An intense research in this field has led to the identification of skyrmions in thin-film and multilayer structures in these heterostrutres skyrmions are able to survive at room temperature and can be manipulated by electrical currents. Utilizing interlayer magnetic exchange bias with synthetic antiferromagnet with can be used to isolated antiferromagnetic skyrmions at room temperature. The development of skyrmion-based topological spintronics holds promise for applications in the writing, processing and reading functionalities at room temperature and can be extended further to all-electrical manipulation spintronics.

Keywords: Skyrmions, thin films, interfaces, magnetism, multilayers

1. Introduction

Skyrmions in magnetic materials are topological spin textures that are stabilized in various types of magnetic interactions. Non-centrosymmetric bulk magnetic materials exhibit chiral, polar, or D_{2d} symmetry which provide a good area to study the topological spin structures and emergent electromagnetic responses arising from them [1, 2]. In these materials without the presence of inversion symmetry, Dzyaloshinskii–Moriya interaction (DMI) can gradually twists the ferromagnetic spin arrangement, giving rise to helimagnetic structure in zero field as well as skyrmions in presence of a magnetic field. The stabilization of Skyrmions is not only due to deformation of helical spin structure but the topologically protected state from external perturbations and thus appropriate for carrying information in robust manner. Skyrmions can be different in to three types of as schematically shown in **Figure 1** have been observed experimentally, which were theoretically predicted Such as Bloch, Néel and Anti skyrmions [3].

Bloch-type skyrmions Bloch-type skyrmion is the first category in chiral magnetic materials, where the spins are lying within the tangential plane of the system,

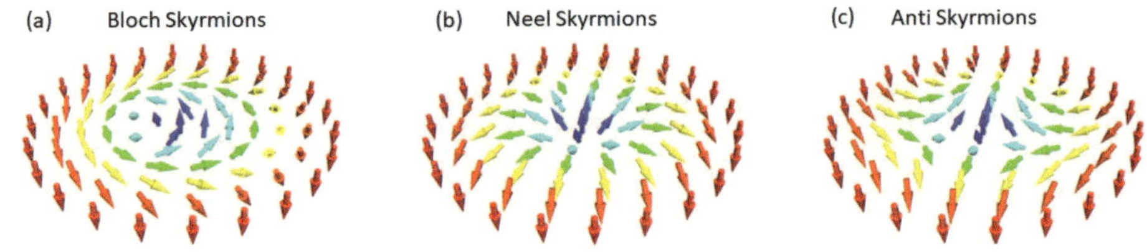

Figure 1.
(a) Bloch-type skyrmion is observed by a transverse helix with an anticlockwise spin rotation, (b) Néel-type magnetic skyrmion is observed by the anticlockwise-rotated magnetization in a spin cycloid and (c) antiskyrmion has a structure of boundary walls that have alternating Bloch and Néel types.

as shown in **Figure 1(a)**. It was first experimentally observed in the A-phase of MnSi with B20-type chiral magnetic structure. The skyrmion structure is described as a superposition of three screw-type helices, by small angle neutron scattering (SANS) [4] and later by Lorentz transmission electron microscopy (LTEM) technique [5]. Later on other bulk systems, in 2012, skyrmions were observed in multiferroic Cu_2OSeO_3 [6]. In 2015, β-Mn type Co-Zn-Mn alloy showed metastable skyrmion at zero magnetic field and room temperature [7], as displayed in **Figure 2**. The formation of the skyrmions was confirmed in the real space by Lorentz transmission electron microscopy (LTEM) along the (110) direction of the sample and it thickness was 150 nm. The skyrmions state was seen at 370 K at the 0.07 T magnetic field and it was persistant till 290 K during the field cooled process. The Skyrmions state was not influenced by the field cooling protocol. The characteristic swirling of the magnetic moments in a skyrmion is observed by the angular color coding in **Figure 2**. In the zero field protocol the magnetic skyrmions are seen also at 350 K and the helical state was seen at 370 K.

Néel-Type skyrmions Néel-type of skyrmion is the second class of skyrmions. It is observed in polar magnets with the spins are lying within a radial plane, as shown in **Figure 1(b)**, and their lattice is described as a superposition of three cycloidal helices. The bulk material of GaV_4S_8 identified to show Néel-type skyrmion which is a polar compound with a Lacunar spinel structure (**Figure 3**) [8].

Antiskyrmions Antiskyrmion is the third type in the skyrmion family as shown in **Figure 1(c)**. It's structure can be understood in terms seeing along the cross-sections of an antiskyrmion with either screw-type helix or cycloidal one, depending on the azimuthal direction of observance. It was discovered in Heusler compounds with D_{2d} crystal symmetry in 2017 and experimental observed using LTEM **Figure 4** [9]. Application of magnetic fields stabilizes the antiskyrmion in the lattice in the wide range temperature including room temperature. The antiskyrmions in thesystem in detail, an under-focused LTEM image of a single antiskyrmion taken at a field of 0.29 T applied parallel to the [001] direction is displayed in **Figure 4(a)**. Most interestingly, it exhibits two bright spots and two dark spots along the [010] and [100] directions, respectively. The modulation of the contrast can be discerned from the appearance of two peaks with a trough in the middle of the line profile taken along [010] (lower inset) and two troughs with a peak in the middle of the line profile taken along [100] (upper inset). A slight distortion of the lattice along [010] might be related to the presence of a small amount of in-plane field due to a slight misorientation (by approximately ±3°) of the sample away from the exact [001] direction. At low in-plane fields, the distortion is mostly in the [010] direction as the helix propagates along [100]. Owing to the tetragonal D2d symmetry, we expect that the antiskyrmion lattices undergo characteristic distortions in oblique fields, because their axis remains locked to the tetragonal axis, and that the antiskyrmions are consequently distorted, with the perfectly radial core shifting from the centres of the cells of the hexagonal densely

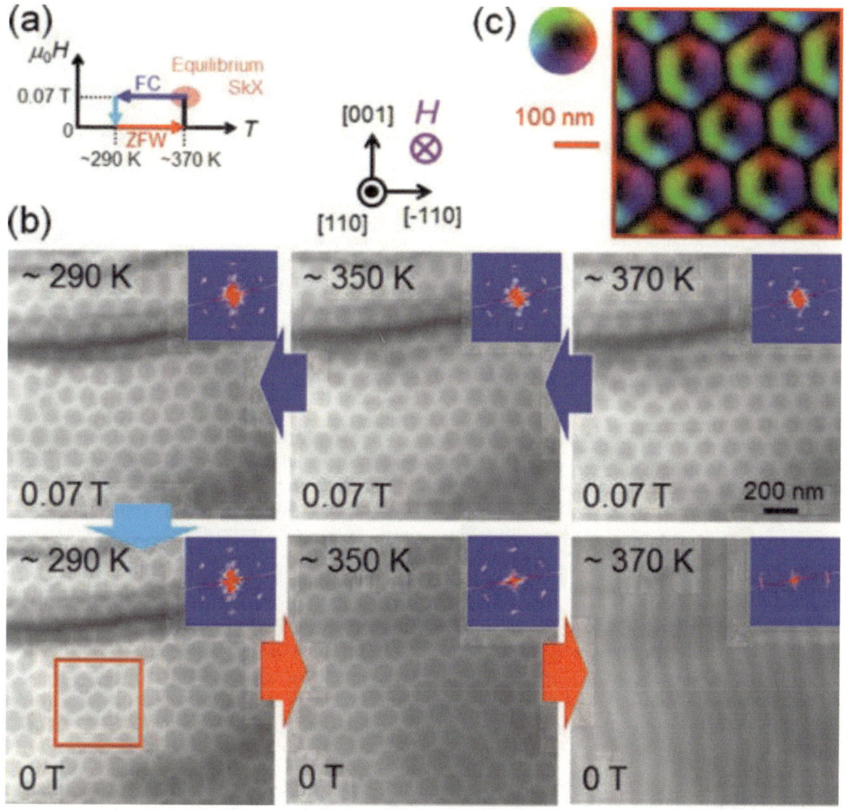

Figure 2.
Schematic illustration of the measurement process. The colors of the arrows correspond to the colors of arrows in (b). (b) Overfocused LTEM images on the (110) plane in the FC process and in the subsequent ZFW process. The defocus values of LTEM images were set to be +28 8μm at 370 K,192 μm at 350 K, and + 96 μm at 290 K, respectively. These LTEM images were taken for nearly the same sample position. The insets of the respective panels show the fast Fourier-transformed patterns of the LTEM images taken over a wider sample area. (c) Distribution of in-plane magnetic moments for the area indicated by the red square in (b) at 290 K and zero magnetic field after the FC, as deduced from a transport-of-intensity equation analysis of overfocused and underfocused LTEM images. The color wheel represents the direction and magnitude of the in-plane magnetization [7].

packed lattice. in **Figure 4(d)**, illustrating the skewed appearance of the antiskyrmions. This effect occurs because the in-plane component of the magnetic field favors and broadens one edge of the circulating magnetization along the field, at the cost of the opposite edge. As can be seen in **Figure 4(e)**, when a magnetic field was applied at an angle of about 20 degrees to [001] by rotating the sample along [110], the antiskyrmions become distorted. We find that the antiskyrmions are stable up to this maximum possible tilt angle. The appearance of double-skyrmion-like features results from the projection of the antiskyrmion tubes that are oriented obliquely to the illuminating electron beam and to the applied field. As the rotation angle decreases (**Figure 4e–g**), the original quadrupole-like internal magnetization distribution of the antiskyrmions is revealed. Strong bright spots can be seen at the upper half of the antiskyrmions for a rotation angle of θ = 8 (**Figure 3g**), whereas they appear at the lower half for θ = 5° (**Figure 4** (h)). In both cases, the antiskyrmion lattice exhibits a large distortion along [010]. Nearly symmetric bright and dark spots in an almost hexagonal lattice can be seen in **Figure 4(i)**, for an applied field of 0.29 T along [001] (within the ±3° limit). A small increase in the field to 0.33 T perturbs the regular arrangement of the antiskyrmions in the lattice **Figure 4** (j). This field corresponds to the stability limit of the equilibrium

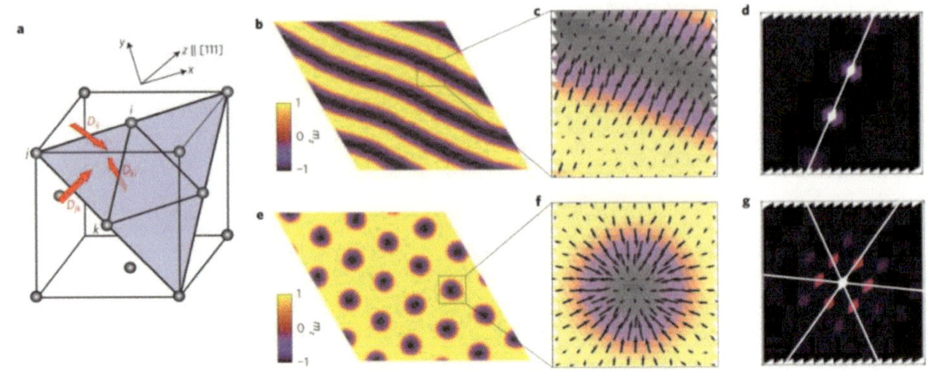

Figure 3.
a, FCC lattice of V4 units, each carrying a spin 1/2, and the orientation of the Dzyaloshinskii–Moriya vectors for bonds on the triangular lattice within the (111) plane (chosen as the xy plane in the calculation). b, Cycloidal spin state obtained for the spin model in equation (1) on the triangular lattice in zero magnetic field. The color coding indicates the out-of-plane components of the spins. c, magnified view of the magnetization configuration for the cycloidal state. The arrows correspond to the in-plane components of the spins at every second site of the triangular lattice. (The remainder of the sites are not shown to reduce the density of the arrows and preserve the clarity of the figure.) d, Bragg peaks (q-vectors) of the cycloidal state in b in reciprocal space. e, SkL state obtained for the spin model in equation (1) (To determine the spin patterns classical Heisenberg model on the triangular lattice was carried out by Monte Carlo technique) on the triangular lattice for B/J⊥ = 0.08 along the z axis. The color coding is the same as in b. f, Magnified view of the magnetization configuration for the SkL state clearly shows the Néel-type domain wall alignment. Note that the magnetization points opposite to the magnetic field in the core region of the skyrmions. (Similarly to c, only every second spin is shown.) g, Bragg peaks of the SkL state in e. The q-vectors of first-order Bragg peaks are located along the ~10~ directions (white lines) in the hard plane, for both the cycloidal and SkL states [8].

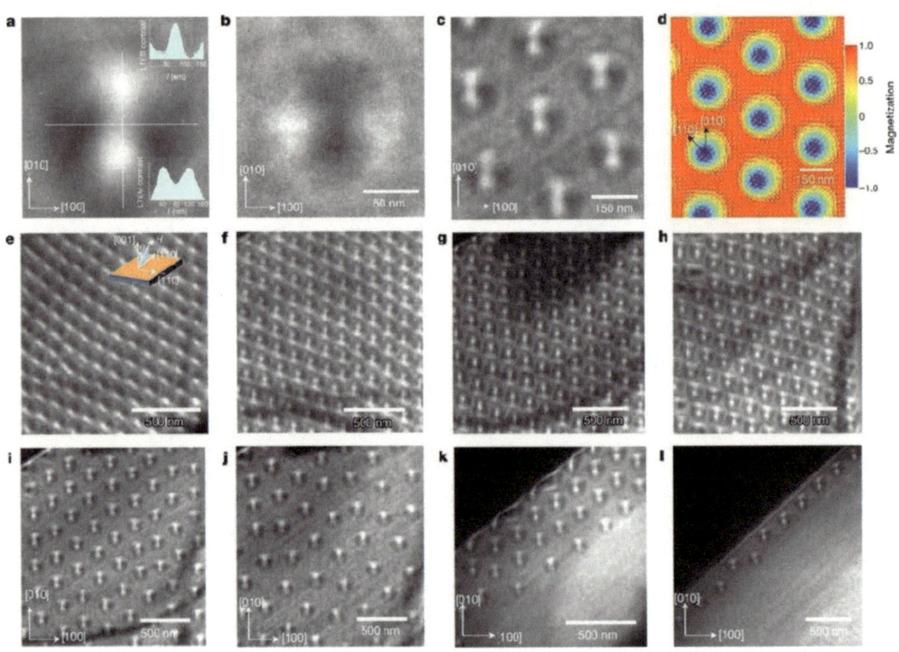

Figure 4.
Under-focused LTEM image of a single antiskyrmion at 300 K and with a field (0.29 T) applied along [001]. The lower and upper insets show the intensity profiles of the in-plane magnetization along [010] and [100], respectively. The corresponding scanned regions are marked by lines. I, distance. b, Over-focused LTEM image of the single antiskyrmion shown in a. c, Under-focused LTEM image showing a hexagonal lattice of antiskyrmions. d, Theoretical simulation of an antiskyrmion lattice in an oblique field. The color represents the magnetization component normal to the sample plane. E–h, under-focused LTEM images of antiskyrmions taken at 300 K and in H = 0.24 T with rotation angles (θ; as shown schematically in the inset of e) of θ = 20° (e), θ= 13° (f), θ = 8° (g) and θ =5° (h). i–l, Under-focused LTEM images of antiskyrmions taken at fields applied along [001] of 0.29 T (i), 0.33 T (j), 0.24 T (k) and 0.49 T (l) [9].

lattice phase. Owing to their topological stability, a large number of antiskyrmions remain as metastable excitations in the homogeneous field-polarized collinear state. For even higher fields, antiskyrmions disappear from the relatively thinner region of the sample and are stabilized only in the thicker region (**Figure 4(k)**). Finally, the antiskyrmion lattice evolves into an array of single antiskyrmions, which disappear for fields above 0.49 T at room temperature (**Figure 4(l)**).

It can be observed that some materials exhibit (anti)skyrmions above room temperature, but continued efforts to has been made in this area to expand the horizon of such materials and necessary understanding of the fundamental physics of the skyrmions is required for their device applications. Further three-dimensional nature of skyrmions in non-centrosymmetric bulk materials can provide unique functionalities where the directionally non-reciprocal transmission of spin excitations can be manipulated. The spin excitations are difficult to observe for interfacial-DMI-based skyrmions in magnetic multilayer systems. **In the following sections, we describe current and future challenges, followed by advances in science and technology to meet them, which are commonly important for all the three types of bulk skyrmions.**

2. Skyrmions in mono layer thin films and multilayer interfaces

From the previous section it can be observed that the magnetic skyrmions were initially identified in single crystals of magnetic materials with a non-centrosymmetric lattice [4, 5], and explained by the existence of DMIs induced by spin–orbit coupling in the absence of inversion symmetry in the crystal lattice.

2.1 Monolayer epitaxial thin films

Skyrmions were observed in epitaxial ultrathin magnetic films grown on heavy metals, which are subject to gaint DMIs induced at the interface that breaks inversion symmetry and the strong spin–orbit coupling with neighboring heavy metal. The first investigated systems in this class were Fe monolayers and PdFe bilayers on Ir(111), huge DMI was seen at the interface of Fe/Ir(111) [10]. Skyrmions observed in these systems are extremely small and was extending for only a few lattice parameters (5(c)). However, the skyrmions are stabilization in large magnetic fields of 1 T and low temperatures and at low temperature of 30 K. Moreover, the skyrmion lattice ground state of an Fe monolayer on Ir(111) does not allow the specific properties of individual skyrmions to be exploited. In PdFe bilayers epitaxially grown on Ir(111), spin spirals are observed at low field with an applied field of about 1 T induces a ferromagnetic state embedding individual metastable skyrmions transition to a [11].

The a two-dimensional square lattice of skyrmions was observed on a single atomic length scale. The magnetic ground state of a hexagonal Fe film of one-atomic-layer thickness on the Ir(111) surface. The real space image of skrmion was imaged using the Using spin-polarized scanning tunneling microscopy by which we can directly image the non-collinear spin texture in and demonstrate that it is incommensurate to the underlying atomic lattice [10]. To investigate the magnetic ground the measurements were carried out with and without an external magnetic field using a tip sensitive to the in-plane magnetization component of the sample. The SP-STM image measured with an in-plane magnetized tip is shown **Figure 5a**, shows all three possible rotational magnetic domains due to the combination of square magnetic structure and a hexagonal atomic lattice. Using the of the tip with one particular magnetization they were able to catch different components of the

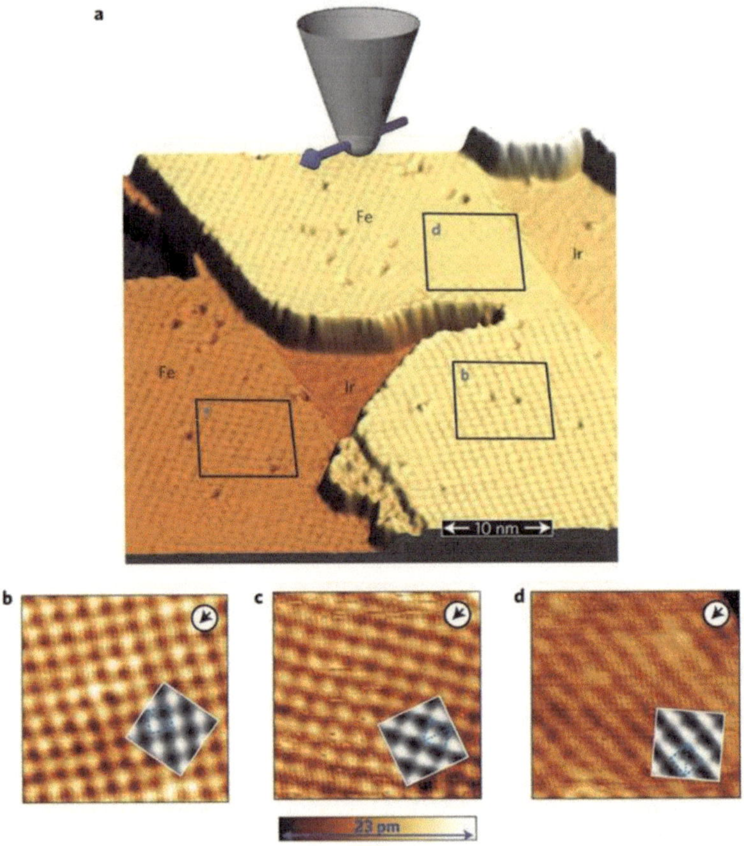

Figure 5.
(a) Three-dimensional representation of a sample area with all three possible rotational magnetic domains measured with a tip sensitive to the in-plane component of magnetization as shown (Fe-coated W tip, U = +5 mV, I = 0.2 nA, sketched tip magnetization axis inferred from comparison to simulated SP-STM images). B–d, closer views of the three rotational domains indicated by squares in a; the tip magnetization is indicated by the arrows. Insets: The simulations of SP-STM measurement of the nanoskyrmion with this tip magnetization; image size and unit-cell position identical owing to the relation between the three rotational domains the magnetic unit cell is rotated by 120° from domain to domain. Note the color scale applies to b–d only [10].

in-plane sample magnetization of the rotational domains with respect to a unique axis in the magnetic unit cell. A closer view of each magnetic domain is shown in **Figure 5(b–d)**. We can analyze the images by considering that all of them have been measured with an identical magnetic tip and the tips magnetization direction indicated by arrows) while the magnetic unit cell rotates by 120° from domain to domain. The excellent agreement between SP-STM experiment and simulated images in the insets clearly shows that our magnetic structure is in accordance with the nanoskyrmion lattice. Based on SP-STM measurements for the in-plane and out-of-plane magnetization components we can even construct the vector magnetization density of our sample, revealing the characteristic spin structure of the skyrmion lattice. **Figure 5d**, is at the resolution limit of STM which is about one-third of that for the square patterns **Figure 5b,c** and seems to require a very high spin polarization of the tip. However, the observation of this pattern in a simultaneous measurement on three different domains–which has not been possible in a previous measurement–is crucial to unambiguously identify magnetic in-plane sensitivity. By applying an in-plane magnetic field, we can align the tip magnetization direction with respect to the crystallographic directions and experimentally rule out a vortex lattice, which could also explain the measurement of **Figure 5**. The reading and writing of the magnetic domains was demonstrated by using the topological property of the skyrmions [11]. By injecting electrons locally they have we generated the desired skyrmion configuration for the four pinning sites within the area. In the series of difference SP-STM images in **Figure 6(b)** to **(e)**, the skyrmions are annihilated one after the another until no skyrmion is present as shown in **Figure 6(f)**. The skyrmions are then created in a different sequence until the starting part of the configuration is reached again as shown in the **Figure 6(g)** to **(j)**. By sweeping the voltage locally the writing and deleting was done. These measurements have demonstrated that the skyrmions can be manipulated individually and independently in a close proximity to one another.

The chiral magnets with archetypal B20 family like MnSi, $Fe_xCo_{1-x}Si$, FeGe, $Fe_{1-x}Co_xGe$, $Fe_{1-x}Mn_xGe$ and MnGe [12, 13]. These B20 silicides and germanides lattice matches well with Si(111) and can be grown epitaxially on the technologically relevant substrates like MgO and SiC(0001). High quality films on large substrates area create the possibility to explore the influence of anisotropy and finite size effects on chiral magnets. Anisotropy provides an important mechanism for increasing the skyrmion stability, which is enhanced along the easy direction, or in an easy-plane, but is reduced along a hard-axis or in a hard-plane relative to the competing cone phase The micromagnetic calculations for MnSi [12], shown in **Figure 7(a)**, indicate that a skyrmion in grating are more energetically favorable than the helicoids. The spin asymmetry scattering expected from the one dimenaional helicoidal modulation does not agree with the measured asymmetry. In the sample interfacial twists at the chiral grain boundaries repel skyrmions and restrict their elliptic distortion below the strip-out field. The energy barriers associated with the film interfaces and grain boundaries lead to metastable structures consisting of helicoid as shown in **Figure 7(b)** [13]. The repulsive force imposed by the surface twist of helimagnet creates lateral confining potential for moving skrmions along the centre of as shown in **Figure 7(c)** [14] Finite size effects arise from the presence of DMI in the zero-torque boundary conditions, which creates surface states that decay into the bulk of the film on a length scale set by the helical

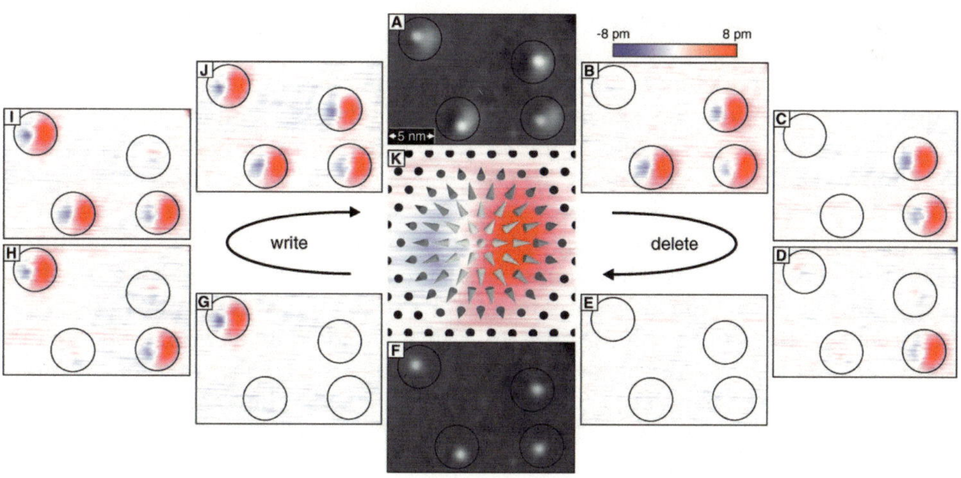

Figure 6.
(a) Constant-current image of a sample region with four defects each hosting a skyrmion marked by a circle containing 270 surface atoms (V = +250 mV, I = 1 nA, B = +3.25 T, T = 4.2 K, magnetically in-plane tip. (b) to (e) Sequence of difference SP-STM images with respect to (f) showing the selective erasing of all four skyrmions using local voltage sweeps. (f) The sample area without skyrmions in a constant-current mode and (g) to (j) their successive rewriting with the difference images. (k) Is the schematic spin configuration with distances twice the atomic lattice, superimposed on the experimental data [10].

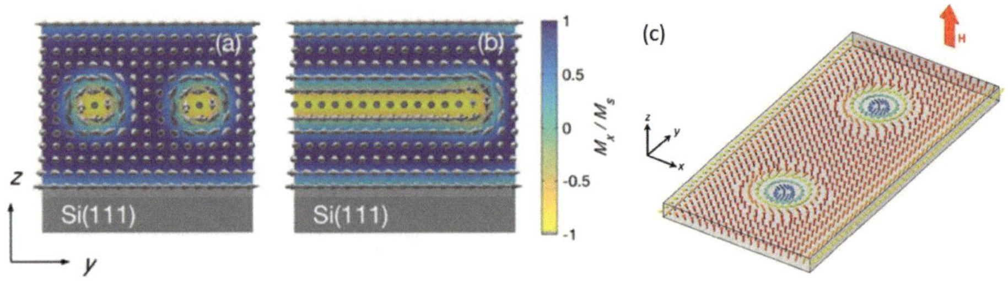

Figure 7.
The calculation of the magnetization M on the y-z plane of (a) 26.4-nm-thick MnSi film where the color plot corresponds to mx. (a) Skyrmiongrating for a field of μ_0 H = 0.5 T. (b) Half of a metastablehelicoidal structure at a field of μ_0 H = 0.3 T, formed from a skyrmion when the applied field is dropped below the skyrmion elliptic instability. (c) Repulsive chiral surface twists that arise at the edges of a narrow strip of a saturated helimagnet force isolated skyrmions to move along the central line of the strip [12].

wavelength, LD and the applied magnetic field. One-dimensional surface twists first were first observed in polarized neutron reflectometry measurements in the field induced ferromagnetic state of MnSi/Si(111) [12]. These twists produce a confining potential for skyrmions that is crucial for devices, since film edges remove the skyrmions' topological protection [14]. Micromagnetic calculations show that the surface states produce a cross-over in the magnetic behavior for film thicknesses below approximately 8LD, where skyrmions are stabilized over a large portion.

2.2 Multilayer thin films

In multilayers for the skyrmions are formed by stacking layers of magnetic and non-magnetic heavy thin metallic films [15, 16]. These multilayer exhibit promising results on the observation of individual skyrmions. Where as in the monolayer a collection of skyrmions are observed. The multilayer approach offers advantages and opportunities for tailoring the skyrmion characteristics. By changing tuning the interactions with different material combination between magnetic metals, heavy metals, ferroelectrics and by varying the number of repetitions in the multilayers.

The multilayered structure combines the interface-driven, out-of-plane magnetic anisotropy and additive DMI at successive interfaces [16, 17]. The room temperature skyrmions have been recently obtained in for example, those formed by 10 repetitions of 0.6 nm of Co sandwiched between 1 nm of Ir and 1 nm of Pt — the additive DMI at the Co/Ir and Co/Pt interfaces [15, 16] induces skyrmions in the Co layers. The skyrmions in successive Co layers are coupled through the ultrathin non-magnetic layers, the large magnetic volume of the resulting along the columns of the thin film of couples. The Monte Carlo simulations for fcc stacking of the Pd overlayer is displayed the low-temperature phase diagram as shown in **Figure 8a**. At zero magnetic field an spin spiral state with a period of about 3 nm is shown in **Figure 8b**. In the lattice the distance between adjacent skyrmions amounts to about 3.3 nm and the system undergoes a second phase transition into the saturated ferromagnetic state for magnetic fields above 17 T. The two phase boundaries metastable state are mixed and appear on cooling down the sample in a finite magnetic field, for example, composed of spin spirals and single skyrmions in **Figure 8c**. For larger fields the isolated skyrmions in a homogeneous ferromagnetic background.

The magnetic properties in spin glasses revealed that chiral magnetic interactions between neighboring spins can also play a key role in metallic systems. The RKKY model was extended to 3-sites to the bilayer systems has been predicted by [18] that a large Dzyaloshinskii–Moriya interaction (DMI) with magnetic film and a

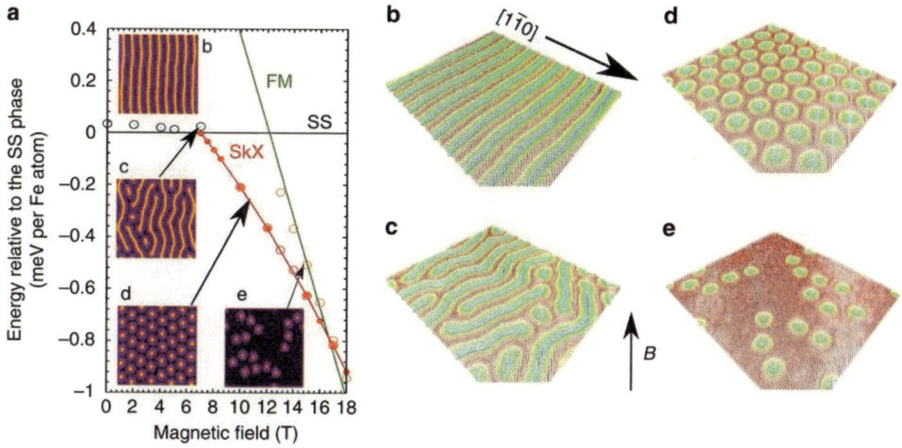

Figure 8.
(a) Phase diagram for fcc stacking of the Pd overlayer on the Fe ML on Ir(111) at low temperatures as a function of a magnetic field applied perpendicular to the film. The energy per Fe atom of the spin spiral (SS), the skyrmion lattice (SkX) and the saturated ferromagnetic (FM) state are given by black, red and green lines, respectively. Open circles indicate the energy of mixed states. Insets show the simulated spin-polarized scanning tunneling microscopic images for an out-of-plane magnetized tip of the spin structures that are displayed in b–e. (b–e) A red color denotes magnetic moments pointing up, that is, in the direction of the magnetic field, while blue spins point into the opposite direction [15].

heavy material with large spin–orbit coupling. The best solution up to now seems to increase the effective magnetic volume by using multilayer stacks composed of multiple repetitions of thin magnetic metal layers separated by heavy metal nonmagnetic layers grown by sputtering deposition. This approach enables the increase of the thermal stability of columnar skyrmions, that are coupled in the successive layers, leading to the recent observation of sub-100 nm skyrmions stable at RT in different multilayer systems [19, 20]. The quasi static XMCD-STXm images

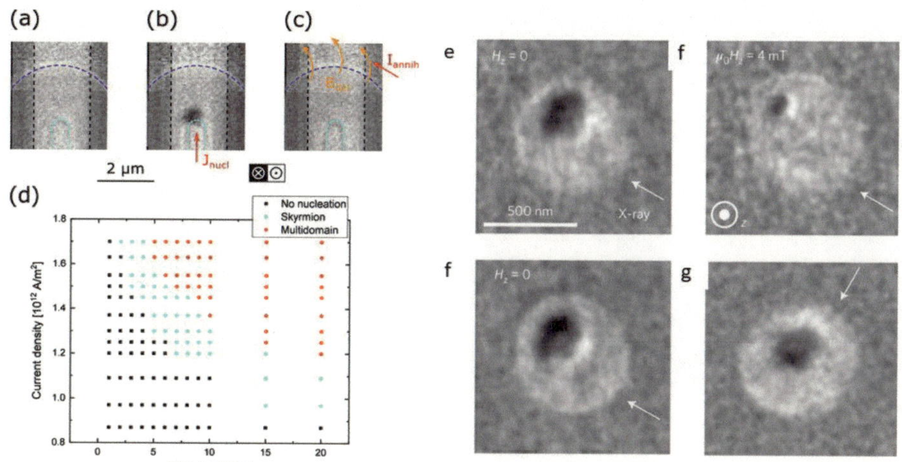

Figure 9.
(a–c) XMCD-STXM images of the current-induced nucleation and field-induced deletion of a magnetic skyrmion in the quasi-static mode, (b) nucleation of a magnetic skyrmion, (c) recovery of the initial magnetic configuration by injecting a current pulse across the microcoil, leading to the generation of an out-of-plane magnetic field pulse and (d) dependence of the current density required to nucleate an isolated magnetic skyrmion on the duration of the current pulse injected (e), XMCD-PEEM image of a magnetic skyrmion in a 630 nm diameter circular dot. (f), XMCD-PEEM image of the same skyrmion during the application of an external magnetic field perpendicular to the film plane μoHz = 4 mT. (g), image taken after b for H_z = 0. (h), XMCD-PEEM image of the skyrmion after rotation of the sample by 90° with respect to the X-ray beam direction. The white arrows indicate the direction of the X-ray beam.

were carried out as shown in **Figure 9(a-c)**, the nucleation and the detection of the protocol employed. The **Figure 9(b)** shows the nucleation of the magnetic skyrmions by the 5 ns current pulse and the recovery if the initial configuration by the microcoil as shown in **Figure 9(c)** and the skyrmion nucleation and their window can be observed from these measurement is shown in **Figure 9(d)**. **Figure 9(e)** shows the XMCD-PEEM image of a magnetic skyrmion with an 630 nm diameter. The larger skyrmion diameter of 190 nm is due to the larger Co thickness. When applying a field μ_0 Hz = 4 mT as shown in **Figure 9(f)** in the direction opposite to the skyrmion magnetization the size of the skyrmion decreases to 70 nm. When the external magnetic field is releasing the initial skyrmion structure is recovered as shown in **Figure 9(g)**. This demonstrates that the skyrmion structure is stable and reversible with respect to perturbations and that the skyrmion diameter can be tuned using frequency. The chiral structure of the skyrmion, we also imaged the skyrmion for an X-ray beam direction rotated by 90° in-plane with respect to the sample as shown in **Figure 9(h)**.

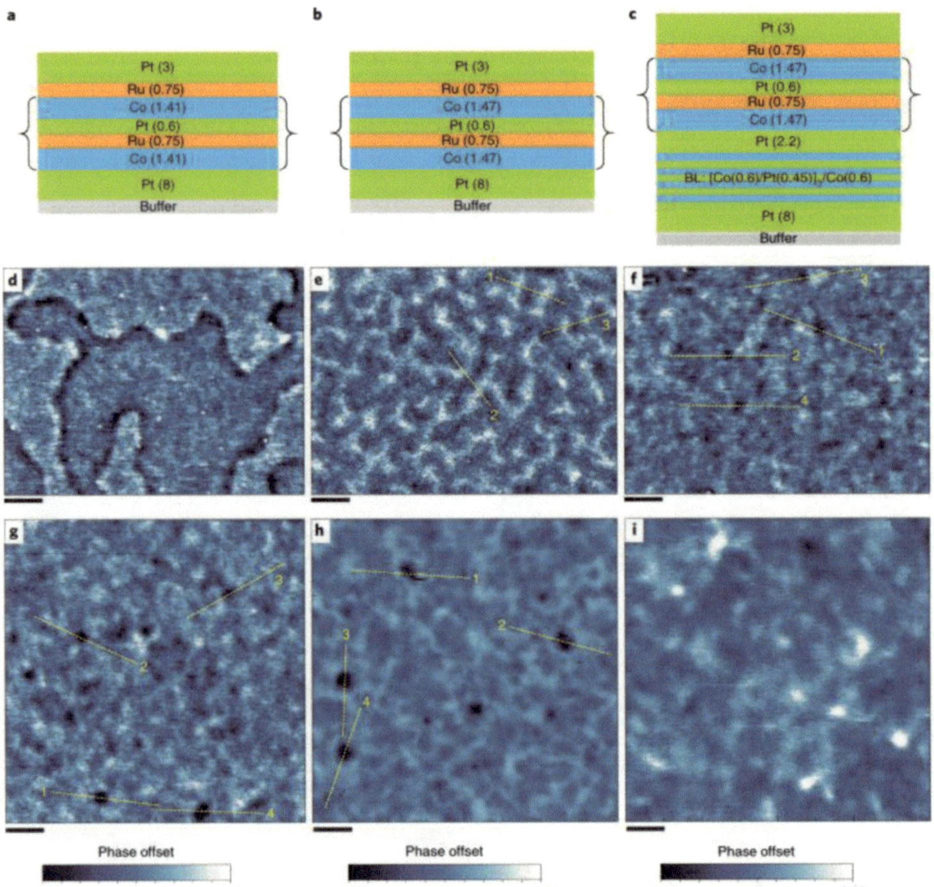

Figure 10.
Multilayer stack for SAF with substantial effective PMA (a), SAF with vanishing effective PMA (b) and BL-SAF system (c). Brackets indicate the SAF system, layer thicknesses are given in nanometers. Note that t_{Co} = 1.41 nm is well below whereas t_{Co} = 1.47 nm is very close to the spin reorientation thickness previously determined. d, MFM image of SAF with effective PMA at 0 mT. e, MFM image of SAF with vanishing effective PMA at 0 mT. f, MFM image of BL-SAF at 0 mT, after saturation of the BL under an external perpendicular field μ_0 H_{ext} = 60 mT. g–i, MFM observations of the BL-SAF system under external applied perpendicular field for μ_0 H_{ext} = 20 mT (g), for μ_0 H_{ext} = 60 mT (h) and for μ_0 H_{ext} = 100 mT (i). Scale bars, 500 nm. All MFM images in d–g share the same color scale shown below g, while MFM images in panels h, i use different scales, all indicating the phase offset in lift mode. Yellow dashed lines in panels e–h indicate line sections of the MFM image.

2.3 Antiferromagnetic materials

The skyrmions hosted in ferromagnets are that has two difficulties that is stabilization of ultrasmall skyrmion due to the dipolar interaction and to stabilize individual skyrmions without using the external magnetic fields [17, 19, 21]. In antiferromagnets, two coupled equivalent magnetic subsystems align antiparallel to each other with no net magnetic moment, thus dipolar fields is absent. This forbid dipolar interactions has attracted researchers extremely interesting to use in skyrmions. MFM image obtained for the in the $4 \times 3\mu m^2$ image as shown in **Figure 10**. The domains in uniform antiferromagnetic configurations with generated the small separation between the two compensating magnetic layers. Perpendicular magnetic ansisotropy and the 1.41 nm is below the 1.47 nm. The spin spiral perodicity constitute the independent measurement of the DMI amplitude for system.

3. Half skyrmions

α Fe_2O_3 is a clasical antiferromagentic materiasl which shows the weak ferromagentic behavior due to the presence of DMI in the material. It crystallizes in the R-3c space group with Néel temperature of 948 K. It also exhibits a Morin transiton in which the spin in-plane at room temperature changes to perpendicular direction [22]. Recently, it was shows that the fabricating the α Fe_2O_3 thin films on the A_2O_3 substrate showed that three possible antiferromagentic (AFM) domain orientations. From the PEEM measurement has visually confirmeed the formation of the Votex state of the Meron which are the half skyrmions and anti-vertex called the Anti-merons as shown in **Figure 11(a)** [23]. When α Fe_2O_3 was exchanged coupled with the ferromagentic Cobalt film of 1 nm thick. The interfacial exchange interaction were able to imprint the AFM on the cobalt thin film as shown in the **Figure 11(b)**. The imprinting of the AFM domains on the cobalt by interfacial exchange with the adjacent AFM vortex/anti-vortex in the α-Fe_2O_3 film. **Figure 11(c)** shows the magnetization in the inplabne of the film. **Figure 11(d)** shows the smearing of the Co vector map. They have also able to manupuulate the meron pairs by the small magnetic field of 100 mT in plane to the [110] direction. This proved that the cobalt layer aligns with the α-Fe_2O_3 and with the staggeres magnetization the majority of the merons have seen disappeared.

Recently the Antiferromagnetic half-skyrmions and bimerons at room temperature was observed [24]. In their study the α-Fe_2O_3 and Rh doped Fe_2O_3 thin films were prepared by pulsed laser deposition and on the top of the film platinum (Pt) was coated on the top of it. For the thin film of α-Fe_2O_3 it showed the Morin transion at 240 K and for Rh doped Fe_2O_3 thin film it was 298 K [25]. It was AFM textures gradually shrink to become fine bubbles and persist up to room temperature for both the films. They have carried out the details analysis by the Néel vector maps of the IP orientations as shown in **Figure 12(a-c)**. In **Figure 12(a)** most of the resions are in the out of plane oriented and display in-plane winding, due to the mixing of the Bloch- and Néel-type domain-wall. They also indicates that the small in-plane islands nucleating inside the out of plane matrix have random in-plane orientations consistent with the first-order nature of the Morin transition. When the temperature is near to the morin transition the out of plane regions shrink and in-plane islands widen considerably and merging with each other if in close proximity as shown **Figure 12(b)**. When the temperature is above the morin transition the AFM textures in which spins are predominantly lying along the basal planes, separated by 60 domain wall **Figure 12(c)**. When a region of out of plane spins

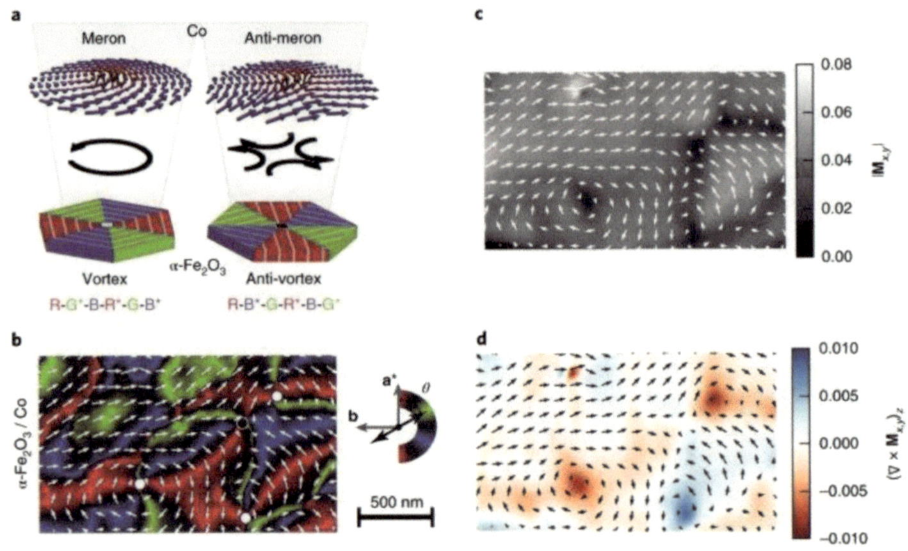

Figure 11.
(a) Shows the Merons (vortices) and anti Merons (anti-vortex) in the α-Fe_2O_3. (b) Domain images of the co and α-Fe_2O_3 thin film, (c) measured magnitude of Co magnetization in the sample plane and (d), z component of the curl of the Co vector map [23].

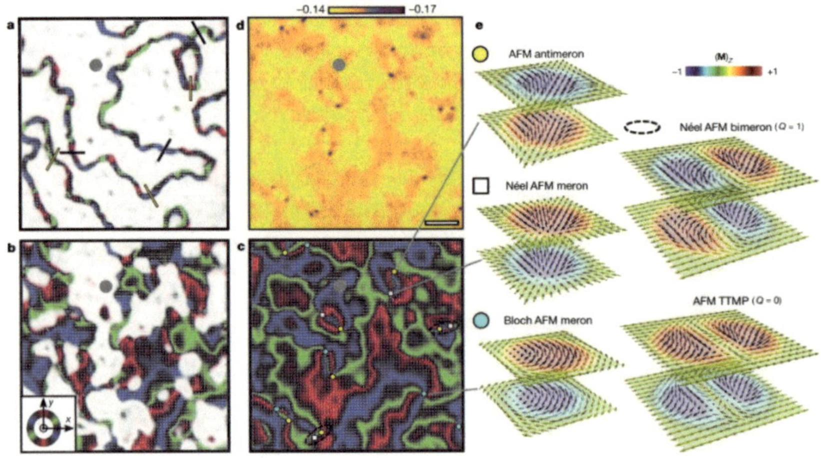

Figure 12.
(a–c) Are the vector-mapped images from the LH-PEEM images afor theα-Fe_2O_3–Pt interface. (a) Temperature less than morin transition, (b) temperature near to morin transition and (c) temperature greater than morin transition. (d) Shows AFM ouf of plane bubbles lying at the core of the corresponding to anti-vertices in 1 μm scale bar [24].

happens to be encircled by in-plane spins, the out of plane bubble becomes trapped and shrinks to a very small size and does not disappear because it is topologically forbidden from unwinding completely into the plane **Figure 12(d)**.

4. Conclusion

From this chapter the various presence of the skyrmions in the multilayer, monolayer and non-centrosymmetric layer have been discussed in detail and the finally the recent advances in the observation of the half skyrmion were also discussed. For the device application thin film heterostructures are very important. This will lead to the future skyrmion based devices for the manipulation by optical, current, electric field and also by the spin orbit torque devices. So that the infomation can be used for the racetrack based memory devices and logic implementation.

Acknowledgements

The author wants to acknowledge IIT Kharapur for SRF funding.

Conflict of interest

The authors declare no conflict of interest.

Author details

Rajesh Kumar Rajagopal
School of Nanoscience and Technology, Indian Institute of Technology Kharagpur, Kharagpur, West Bengal, India

*Address all correspondence to: r.g.rajeshkumarr@gmail.com

References

[1] Bogdanov A and Hubert A 1994 J. Magn. Magn. Mat. 138 255 DOI: 10.1016/0304-8853(94)90046-9

[2] Kanazawa N, Seki S and Tokura Y 2017 Adv. Mater. 29 1603227 DOI: 10.1002/adma.201603227

[3] Bogdanov A and Yablonskii D 1989 Sov. Phys. JETP 68 101

[4] Mühlbauer S, Binz B, Jonietz F, Pfleiderer C, Rosch A, Neubauer A, Georgii R and Böni P 2009 Science 323 915 DOI: 10.1126/science.1166767

[5] Yu X Z, Onose Y, Kanazawa N, Park J H, Han J H, Matsui Y, Nagaosa N and Tokura Y 2010 Nature 465 901 DOI: 10.1038/nature09124

[6] Seki S, Yu X Z, Ishiwata S and Tokura Y 2012 Science 336 198 DOI: 10.1126/science.1214143

[7] Kosuke Karube, Jonathan S. White, Daisuke Morikawa, Charles D. Dewhurst, Robert Cubitt, Akiko Kikkawa, Xiuzhen Yu, Yusuke Tokunaga, Taka-hisa Arima, Henrik M. Ronnow, Yoshinori Tokura and Yasujiro Taguch, Science Advances 2018: 4, 9, 7043. DOI: 10.1126/sciadv.aar7043

[8] Kézsmárki, S. Bordács, P. Milde, E. Neuber, L. M. Eng, J. S. White, H. M. Rønnow, C. D. Dewhurst, M. Mochizuki, K. Yanai, H. Nakamura, D. Ehlers, V. Tsurkan and A. Loidl 2015 Nat. Mater. 14 1116 DOI: 10.1038/nmat4402

[9] Ajaya K. Nayak, Vivek Kumar, Tianping Ma, Peter Werner, Eckhard Pippel, Roshnee Sahoo, Franoise Damay, Ulrich K. Rößler, Claudia Felser and Stuart S. P. Parkin 2017 Nature 548 561 DOI: 10.1038/nature23466

[10] Stefan Heinze, Kirsten von Bergmann, Matthias Menzel, Jens Brede, André Kubetzka, Roland Wiesendanger, Gustav Bihlmayer and Stefan Blügel Nature Physics volume 7, 713–718, 2011 DOI: 10.1038/nphys2045

[11] Niklas Romming, Christian Hanneken, Matthias Menzel, Jessica E. Bickel*, Boris Wolter, Kirsten von Bergmann+, André Kubetzka+, Roland Wiesendanger. Writing and deleting single magnetic skyrmions. Science 341, 636–639 (2013) DOI: 10.1126/science.1240573

[12] S. A. Meynell, M. N. Wilson, K. L. Krycka, B. J. Kirby, H. Fritzsche, and T. L. Monchesky Phys. Rev. B 96, 054402, 2017 DOI:10.1103/PhysRevB.96.054402

[13] Charles S. Spencer, Jacob Gayles, Nicholas A. Porter, Satoshi Sugimoto, Zabeada Aslam, Christian J. Kinane, Timothy R. Charlton, Frank Freimuth, Stanislav Chadov, Sean Langridge, Jairo Sinova, Claudia Felser, Stefan Blügel, Yuriy Mokrousov, and Christopher H. Marrows 2018 Phys. Rev. B 97 214406 DOI: 10.1103/PhysRevB.97.214406.

[14] Meynell S A, Wilson M N, Fritzsche H, Bogdanov A N and Monchesky T L 2014 Phys. Rev. B 90 014406 DOI: 10.1103/PhysRevB.90.014406

[15] Dupé, B., Hoffmann, M., Paillard, C. and Heinze, S. Nat. Commun. 5, 4030 (2014) DOI: 10.1038/ncomms5030

[16] Hongxin Yang, André Thiaville, Stanislas Rohart, Albert Fert, and Mairbek Chshiev. Phys. Rev. Lett. 115, 267210 (2015) DOI: 10.1103/PhysRevLett.115.267210

[17] C. Moreau-Luchaire, C. Moutafis, N. Reyren, J. Sampaio, C. A. F. Vaz, N. Van Horne, K. Bouzehouane, K. Garcia, C. Deranlot, P. Warnicke, P. Wohlhüter, J.-M. George, M. Weigand, J. Raabe, V. Cros and A. Fert. Nature

Nanotechnology 11,444–448 (2016) DOI: 10.1038/nnano.2015.313

[18] Swekis P, Markou A, Kriegner D, Gayles J, Schlitz R, Schnelle W, Goennenwein S T B and Felser C 2019 Phys. Rev. Mater. 3 013001 DOI: 10.1103/PhysRevMaterials.3.013001

[19] Olivier Boulle, Jan Vogel, Hongxin Yang, Stefania Pizzini, Dayane de Souza Chaves, Andrea Locatelli, Tevfik Onur Menteş, Alessandro Sala, Liliana D. Buda-Prejbeanu, Olivier Klein1, Mohamed Belmeguenai, Yves Roussigné, Andrey Stashkevich, Salim Mourad Chérif, Lucia Aballe, Michael Foerster, Mairbek Chshiev, Stéphane Auffret, Ioan Mihai Miron and Gilles Gaudin Nature Nanotechnology. 11, 449–454 (2016) DOI: 10.1038/NNANO.2015.315.

[20] Fert A 1990 Mater. Sci. Forum 59-60 439 DOI:10.4028/www.scientific.net/MSF.59-60.439

[21] Seonghoon Woo, Kai Litzius, Benjamin Krüger, Mi-Young Im, Lucas Caretta, Kornel Richter, Maxwell Mann, Andrea Krone, Robert M. Reeve, Markus Weigand, Parnika Agrawal, Ivan Lemesh, Mohamad-Assaad Mawass, Peter Fischer, Mathias Kläui and Geoffrey S. D. Beach Nature Materials 19, 34–42 (2020) DOI: 10.1038/s41563-019-0468-3

[22] Morin, F. J Electrical Properties of α Fe_2O_3 and α Fe_2O_3 Containing Titanium, Phys. Rev., 83, 5, 1005-1010, 1951, DOI: 10.1103/PhysRev.83.1005. url = https://link.aps.org/doi/10.1103/PhysRev.83.1005

[23] F. P. Chmiel, N. Waterfield Price, R. D. Johnson, A. D. Lamirand, J. Schad, G. van der Laan, D. T. Harris, J. Irwin, M. S. Rzchowski, C.-B. Eom and P. G. Radaelli. Observation of magnetic vortex pairs at room temperature in a planar α-Fe2O3/Co heterostructure. Nature Mater 17, 581–585 (2018). DOI: 10.1038/s41563-018-0101-x

[24] Hariom Jani, Jheng-Cyuan Lin, Jiahao Chen, Jack Harrison, Francesco Maccherozzi, Jonathon Schad, Saurav Prakash, Chang-Beom Eom, A. Ariando, T. Venkatesan and Paolo G. Radaelli. Antiferromagnetic half-skyrmions and bimerons at room temperature. Nature 590, 74–79 (2021). DOI: 10.1038/s41586-021-03219-6

[25] Coey, J. M. D. and Sawatzky, G. A. A study of hyperfine interactions in the system $(FeRh_x)_2O_3$ using the Mössbauer effect (bonding parameters). J. Phys. C Solid State Phys. 4, 2386 (1971). DOI: 10.1088/0022-3719/4/15/025.

Chapter 3

Stabilizing Zero-Field Skyrmions at Room-Temperature in Perpendicularly Magnetized Multilayers

Jeovani Brandão, Marcos Vinicius Puydinger dos Santos and Fanny Béron

Abstract

Magnetic skyrmions are twirling spin structures observed in bulk, thin films, and multilayers with several features for both fundamental physics understanding and spintronic applications, i.e., nanoscale size, efficient transport under electrical current, and topological protection against defects. However, most magnetic skyrmions have been observed under the assistance of an out-of-plane magnetic field, which may limit their use in magnetic memory technologies. In this chapter, we review and present two recent mechanisms to create zero-field skyrmions at room-temperature in ferromagnetic multilayers. First, by tuning the perpendicular magnetic anisotropy (PMA) and remnant magnetization (near magnetization saturation) in unpatterned symmetric multilayer systems, it was achieved a transition from worm-like domains to isolated skyrmions. Besides, we present how to find stable zero-field skyrmions in arrays of ferrimagnetic discs by tailoring their diameter. Both methods demonstrate a robust route to stabilize zero-field skyrmions at room temperature, thus providing an important contribution to possible applications of these textures in the next generation of skyrmionics devices.

Keywords: multilayers, room-temperature, stability, skyrmions, zero-field

1. Introduction

Chiral magnets [1–8] have been investigated intensively in the last years due to their fascinating physical concepts and potential use in applications including sensors [9–11], logic [12–14], and magnetic memory devices [15, 16]. Among different magnetic structures, domain walls [17], vortices [18], and skyrmions [19] have been investigated both theoretically and experimentally. They exhibit remarkable static and dynamic behaviors under external stimulation such as magnetic [20] and electrical fields [21], thermal effects [22], and electrical currents [23]. More specifically, magnetic skyrmions, i.e., small-sized non-collinear chiral spin textures [24], have been attracted much attention as the favorable candidates for information storage in different technological devices, as examples, racetrack memories designed as nanowires [25] and nano-oscillators [26] based on confined nano-discs [27]. This huge interest in manipulating skyrmions lies in their special properties including topological protection [28], efficient transport under low electric current density [29], and reduced (below 100 nm) size [30].

Experimentally, these non-collinear spin structures were firstly observed a little more than one decade ago in bulk systems [31], but at low temperatures (around 30 K) and under magnetic fields (around 0.2 T) [32, 33]. To achieve skyrmions at ambient conditions, perpendicularly magnetized multilayers made of nano-thick magnetic and non-magnetic materials have been used as a platform with suitable parameters to host skyrmions at room-temperature [20]. One important parameter is the Dzyaloshinskii-Moriya interaction (DMI) induced by spin orbit-coupling [34]. In the case of multilayers, it arises due to the interfacial nature between the atoms of the heavy metal (HM)\ferromagnetic (FM) layers [20]. The DMI is responsible for the chirality of the magnetic structure, i.e., the direction in which the spins rotate in the texture, giving rise to the topological skyrmion stability [6].

For multilayers made of thin films, such as Pt/Co/Ta, Pt/Co/Ir, Pt/Co/W, and other combinations, skyrmions have been observed in most cases under applied external magnetic fields [20, 35, 36], while at zero magnetic field the magnetic ground state favors the formation of spin spirals or labyrinthine magnetic configurations [37]. Under the assistance of a moderate magnetic field, the magnetic stripes shrink into small circular structures [20, 35], leading to the creation of magnetic skyrmions. This is possible thanks to the DMI that acts to protect the skyrmion against collapsing into a uniform magnetic state [20]. Moreover, the skyrmion size depends on the magnetic field strength, usually being inversely proportional to the applied field [20, 38].

This mechanism for creating skyrmions in multilayers under the aid of a magnetic field opened an avenue for controlling their fundamental physical properties such as size and density. On the other hand, the usage of skyrmions in the next generation of magnetic memory devices will depend on the achievement of skyrmions without the support of external stimuli. In this sense, the research for efficient methods to establish zero-field skyrmions at room-temperature yields interesting findings. For instance, zero-field skyrmions have been observed at room-temperature in lithographically shaped squares [39], through an exchange bias-field created at the interface of antiferromagnetic/ferromagnetic-based structures [40], and by their direct writing using magnetic force microscopy (MFM) tips [41], X-rays [42], and electron beam [43].

In this chapter, we report on two current processes for stabilizing zero-field magnetic skyrmions in ferro- and ferrimagnetic multilayers at ambient conditions. The different approaches driving the formation of zero-field skyrmions consist of tuning the magnetic properties such as perpendicular magnetic anisotropy (PMA) and DMI in unpatterned symmetric multilayers, as well as by tailoring the geometric size of circular discs. The zero-field skyrmions were imaged by magnetic force microscopy at room-temperature. Both mechanisms establish efficient strategies to create zero-field skyrmions at ambient conditions, which is an important advance in the development of skyrmionics devices.

2. Zero-field Skyrmions in Unpatterned symmetric multilayers

Skyrmions at room-temperature are created in perpendicularly magnetized multilayers made of a magnetic material sandwiched by two different heavy metal layers, HM1 and HM2, which yield sizeable DMI values to topologically protect and stabilize skyrmions under the assistance of a magnetic field [20]. Therefore, most previous works have focused on asymmetric multilayers, which, at zero field and under the competition of different energies (exchange, anisotropy, dipolar, DMI), can lead to the labyrinthine domain structures. These domain structures turn into skyrmions in the presence of a magnetic field below the saturation magnetization [20, 35, 36].

Alternatively, symmetric multilayers made of the same heavy metal layers, for instance Pt/Co/Pt and Pd/Co/Pd, can also be used to create skyrmions [44]. It has been demonstrated that for polycrystalline symmetric Pt/Co/Pt tri-layers deposited by magnetron sputtering, a low but non-negligible DMI is measurable in the system [45]. More recently, a strong DMI has been observed in symmetric Pd/Co superlattices grown by molecular beam epitaxial (MBE) technique [46]. This is due to the differences in the crystalline structures and inhomogeneities at the interfaces of the HM/FM and FM/HM layers, which yield to the formation of asymmetric interfaces. Therefore, a DMI emerges in these irregular interfaces and opens the possibility for exploring chiral magnetic textures in symmetric systems [47].

In this sense, in order to find zero-field skyrmions at room-temperature, multilayers of (Pd/Co/Pd)x15 were fabricated by magnetron sputtering technique, where the Pd thickness was fixed at 2 nm and the Co layer varied from 0.2 to 0.8 nm. **Figure 1** shows the schematic representation of the multilayer Pd/Co/Pd grown on Si/SiO_2 substrate.

2.1 Magnetic properties

Hysteresis loops acquired under out-of-plane magnetic field for different Co thicknesses show that the reversal magnetization behavior depends on the Co thickness (**Figure 2**). At 0.8 nm-thick Co (down blue triangles), the hysteresis loop exhibits a very low remnant magnetization, while its saturation occurs beyond 150 mT, which is a larger magnetic field compared to the ones obtained for thicknesses below 0.8 nm. It essentially means that for this specific thickness, the in-plane anisotropy plays an important role in the magnetization reversal. It has already been demonstrated that above this thickness (0.8 nm thick Co), the magnetization of a similar multilayered system lies in the sample plane [48].

Moreover, by reducing the Co layer thickness to 0.6 nm, the hysteresis loop turns into a tail-like shape with reduced saturation magnetic field (up green triangles), but no significant change is observed for the remnant magnetization. This tendency is altered significantly when the Co thickness is reduced to 0.4 (red circles) and 0.2 nm (black squares). The square format of the magnetization curve is characterized by a reduction of the saturation magnetic field, as well as a notable enhancement of the remnant magnetization (near to the saturation magnetization),

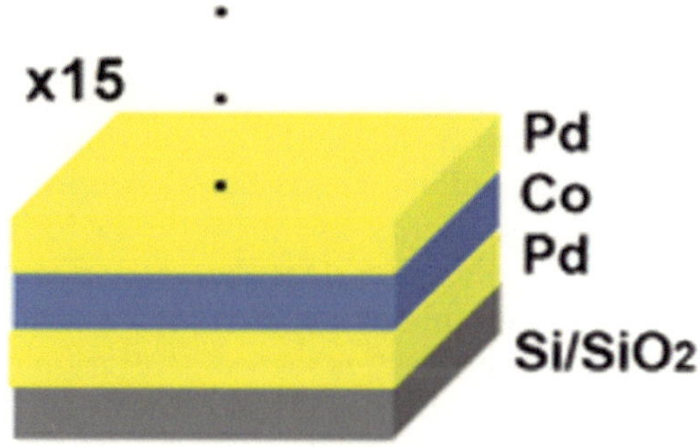

Figure 1.
Stack of the symmetric multilayer produced by magnetron sputtering. Figure adapted from reference [47].

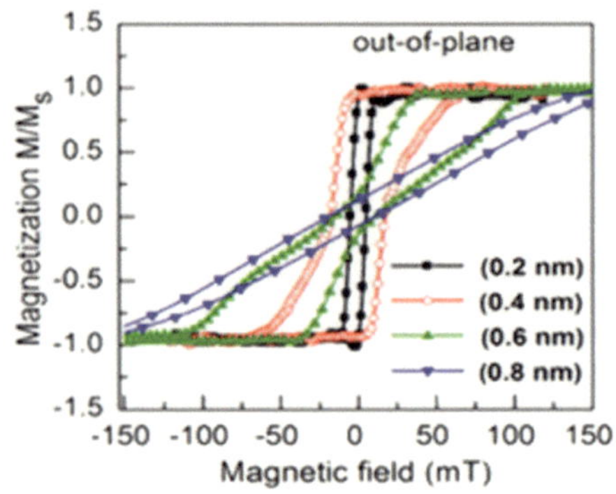

Figure 2.
Out-of-plane hysteresis curves for different Co thicknesses. Figure adapted from reference [47].

indicating that the (PMA) is higher in this thickness range. It yields a thickness-dependency modulation of the remnant magnetization.

Consequently, both the tunable remnant magnetization and the PMA as a function of the thickness play an important role in the resulting magnetic domain patterns at zero magnetic field and room-temperature, as shown in **Figure 3**. The blue frame (0.8 nm) exhibits the formation of small periodic out-of-plane magnetic domains, while the green frame (0.6 nm), of the well-known worm-like domain pattern in such perpendicularly magnetized systems. The domain pattern transition occurs at Co-thick 0.4 nm (red frame), where the MFM image shows long separated magnetic stripes with a few circular domains in between. This is a preliminary indication of the magnetic skyrmion formation in the symmetric multilayer.

The formation of these round structures in symmetric multilayers is further confirmed for the 0.2 nm-thick Co film (black frame), in which the magnetic domain pattern is given by isolated skyrmions at room-temperature and zero magnetic field. This result shows that by carefully tuning the remnant magnetization and PMA of the multilayer, isolated skyrmions can be observed without the need of applying a magnetic field. In other words, increasing the PMA and the remnant magnetization can yield the formation of field-free skyrmions at ambient conditions. Notably, these skyrmions are created only by tuning the multilayer thickness, rather than through any geometric confinement or writing methodologies. This represents a straightforward and feasible alternative to stabilize zero-field skyrmions at room-temperature.

2.2 Skyrmion formation reproducibility

The homogeneity of the magnetic skyrmions along the sample surface was also tracked by taking MFM images over different areas of the 0.2 nm-thick Cofilm, as shown in **Figure 4**. The skyrmions are isolated with randomly distributed positions. This result confirms that the skyrmions at zero-field are reproducible on the entire surface of the multilayer rather than from any topographic influence on the different sample regions. Besides, the skyrmion sizes observed in each image are slightly different, which arises from minor inhomogeneities in the sample and different local variations of the multilayer properties such as saturation magnetization (M_s), PMA, and DMI [49].

To explore the features of the stabilized zero-field skyrmions, their size (diameter) and density (number of skyrmions per area) were extracted from the MFM

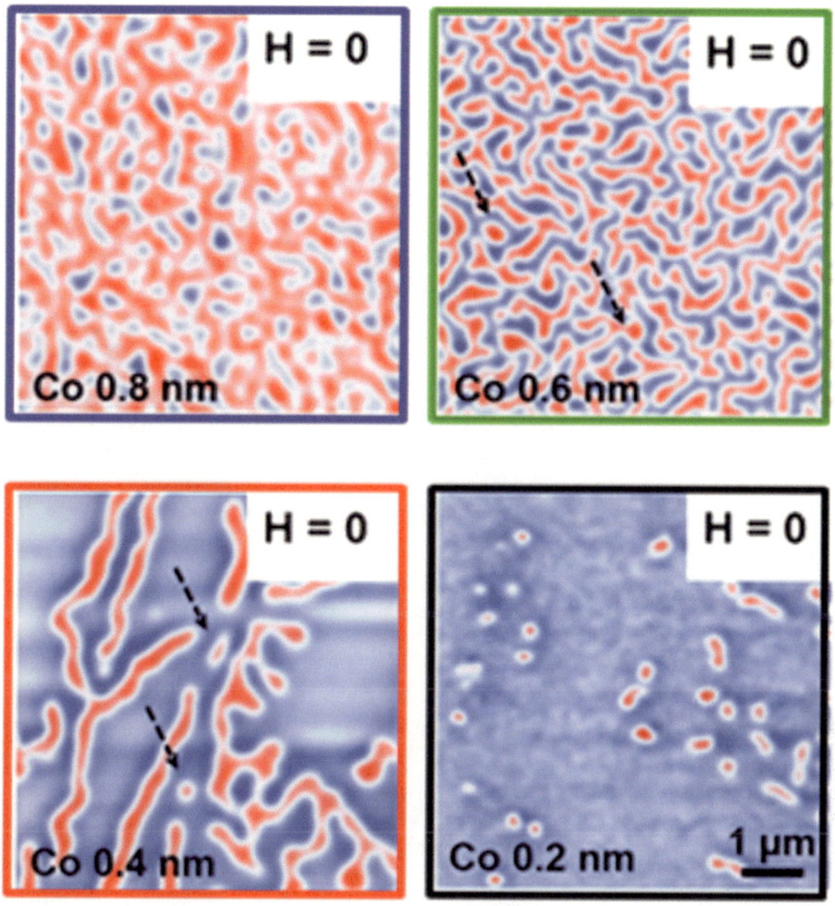

Figure 3.
Magnetic force microscopy (MFM) images of the zero-field domain evolution. Figure adapted from reference [47].

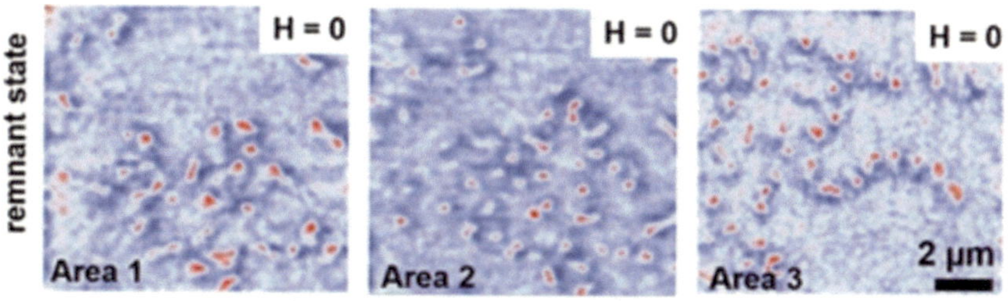

Figure 4.
Magnetic skyrmions observed in different areas of the 0.2 nm-thick Co film at the remnant state. Figure adapted from [47].

images. The skyrmion size was evaluated by performing line scans on the magnetic structures and using the full width at half maximum (FWHM), as shown in **Figure 5(a)** and compiled in a histogram in **Figure 5(b)**, along with the skyrmion density statistics (**Figure 5(c)**). The average size and density of the skyrmions were ~ 165 nm ± 32 nm and 0.25/μm^2. These values agree with previous works where skyrmions were stabilized under applied magnetic field [20, 35], which usually reduces these values [37].

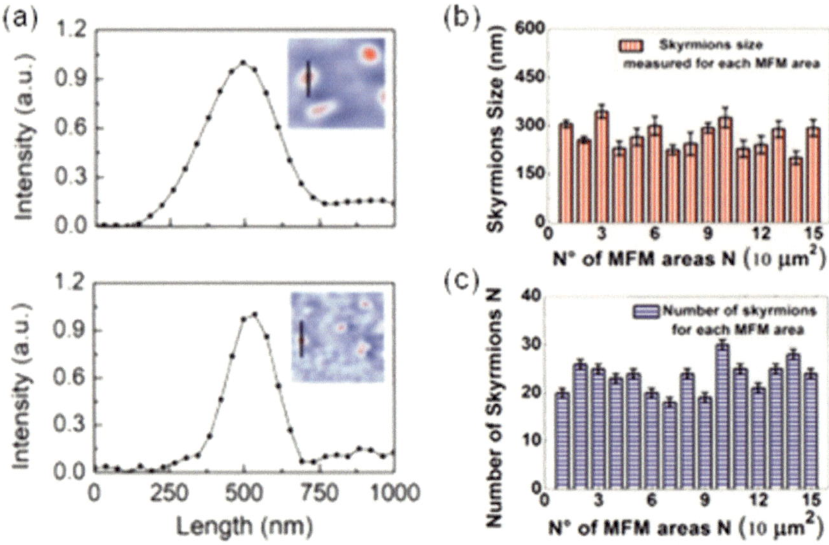

Figure 5.
(a) Line scans performed on skyrmions of different sizes. Histograms of the skyrmion size (b) and number (c) distributions obtained in each MFM image. Figure adapted from [47].

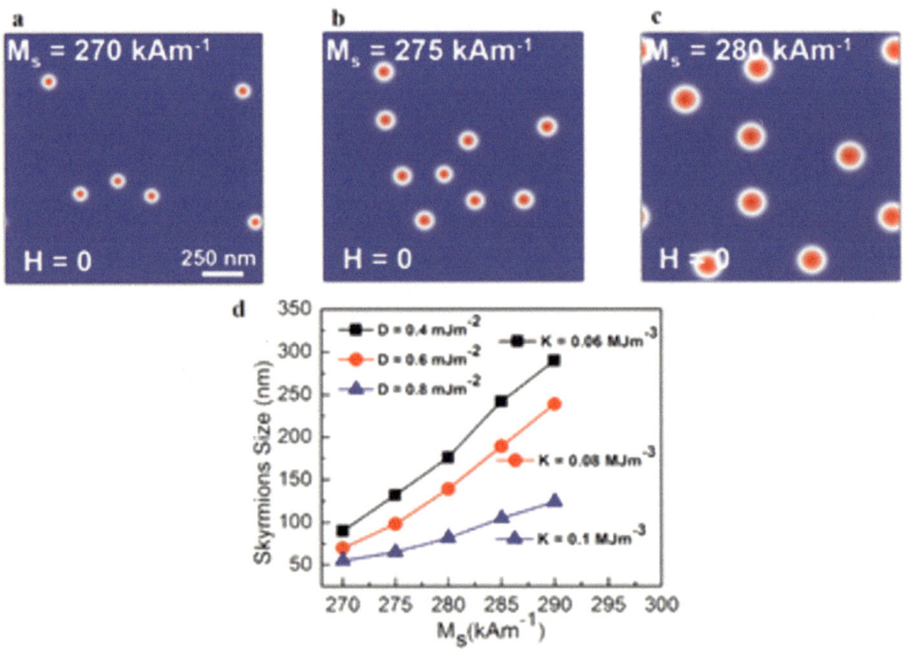

Figure 6.
(a-c) Micromagnetic results of isolated skyrmions stabilized at zero magnetic field. (d) Skyrmion size obtained as a function of M_s for different values of PMA and DMI. Figure taken from [47].

2.3 Micromagnetic simulations

The formation of zero-field skyrmions in unpatterned symmetric multilayers is described better with the assistance of micromagnetic simulations. Mumax3 was the chosen code to carry out the simulations, and the used methods and parameters can be found in [47]. The results are summarized in **Figure 6.**

In **Figure 6(a-c)**, the micromagnetic simulations are representative for an anisotropy K = 0.06 MJm^{-3} and a Dzyaloshinskii-Moriya interaction D = 0.04 mJm^{-2}, while the saturation magnetization M_s was varied. It shows that skyrmions are stabilized at zero-field with a random distribution of their positions, corroborating our experimental findings. Furthermore, by varying D in the micromagnetic simulations and comparing with the size of zero-field skyrmions obtained experimentally, the DMI was estimated to range between 0.4–0.8 mJm^{-2}. More details can be seen in the phase diagrams shown in the next section.

More remarkable, the skyrmion size increases for larger values of M_s as clearly shown in **Figure 6(a-c)**, indicating that this parameter plays an important role on the skyrmion diameter due to the impact of the magnetization stray field upon the magnetic structure size. In addition to the saturation magnetization, the influence of both K and D parameters was evaluated to understand their effects on the skyrmion stability and size. Figure 6 (d) shows that the skyrmion size monotonically increases as a function of M_s for any value of K and D. However, smaller K and D values lead to larger skyrmion size (black squares) compared to higher values (red circles and blue triangles). This important conclusion suggests that the sample saturation magnetization M_s must be reduced, while the PMA and DMI must be increased slightly to reduce the skyrmion average size at zero-field. This can be accomplished by engineering the multilayer structure varying, for instance, the magnetic layer thickness.

2.4 Magnetic domain phase diagrams

A phase diagram was constructed using the micromagnetic simulations to explore and indicate the achievable magnetic ground states as a function of the multilayer magnetic parameters, as can be seen in **Figure** 7.

Three different magnetic ground states were acquired at zero magnetic field, as shown in **Figure 7(a)**: uniform magnetization (dark cyan frame), isolated skyrmions (cyan frame), and mixed phase (purple frame), in which the latter

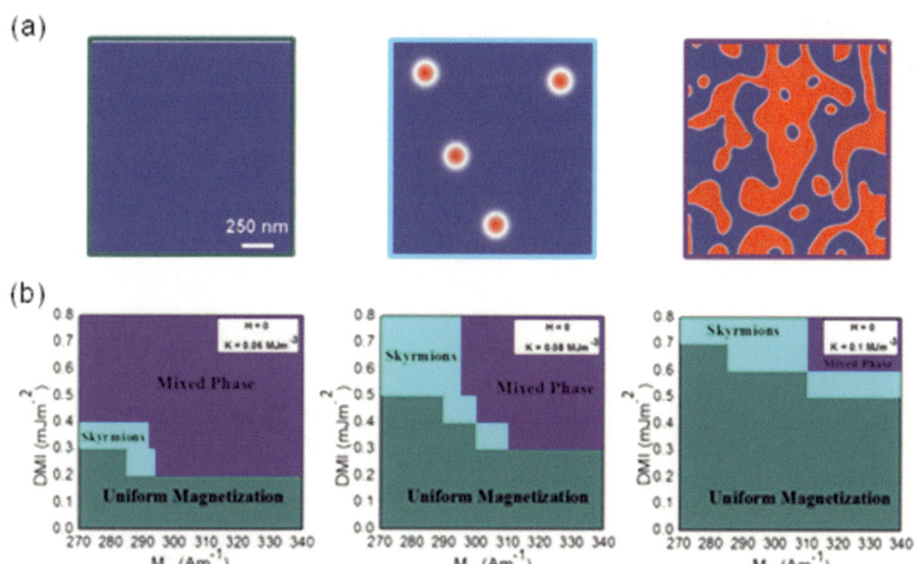

Figure 7.
(a) Possible magnetic ground states. (b) DMI vs. M_s phase diagram for selected values of K. Figure adapted from reference [47].

means that both skyrmions and magnetic stripes can coexist. The phase diagrams show that for low DMI values a uniform magnetization is the preferable ground state along the entire range of saturation magnetization M_s (**Figure 7(b)**). For DMI values above 0.4 mJm^{-2}, isolated skyrmions take place, while the area that covers this magnetic state in the phase diagram expands as the saturation magnetization M_s and K values increase. On the other hand, the mixed phase exhibits an opposite behavior with a reduced area in the phase diagram when M_s and K enhances. It demonstrates that at zero magnetic field, the stability of isolated skyrmions will strongly depend on the system magnetic parameters, mainly M_s and K. Moreover, a DMI non-null value is required to stabilize isolated zero-field skyrmions, which has been achieved for both symmetric multilayers grown by MBE and magnetron sputtering techniques. Therefore, symmetric multilayers that present moderate values of DMI below 1 mJm^{-2}, with high remnant magnetization and PMA, can be strategically used to stabilize skyrmions in the absence of external magnetic field.

3. Zero-field Skyrmions in Ferrimagnetic multilayers

Ferromagnetic skyrmions have been shown remarkable phenomena under external excitation, which have helped to understand their fascinating properties. For instance, when driven by electrical current, ferromagnetic skyrmions experience a transverse motion with respect to the current direction, known as skyrmion Hall effect, which can lead to limiting their use in magnetic memory applications [50]. To overcome this issue, skyrmions have been researched in systems with antiferromagnetic coupling. It has been realized that skyrmions coupled with antiparallel magnetic moments, or, in other words, with opposite topological charges, may be used to suppress the undesirable skyrmion Hall effect when driven by electrical currents [51, 52].

In this context, materials with ferrimagnetic ordering are attractive candidates to host coupled skyrmions with opposite topological charges since they can reduce the skyrmion Hall effect due to the balance of the Magnus force acting on each skyrmion [53, 54]. In this section, we show how to observe skyrmions in lithographically shaped discs made of ferrimagnetic CoGd/Pt-based multilayer by tailoring the size of the disc diameter. For this purpose, unpatterned and patterned multilayers were grown by magnetron sputtering to image the evolution of the magnetic domain structures. **Figure 8** shows the schematic representation of the multilayer in the unpatterned and patterned (discs) geometries. The CoGd alloy was grown by using two independent Co and Gd targets in a co-sputtering deposition. The Pt/CoGd/Pt tri-layer was repeated 15 times and 2 nm Pt under and over layers were deposited as the buffer and protective oxidation, respectively. The CoGd thickness was fixed to 1.8 nm, and both unpatterned and patterned discs were deposited during the same

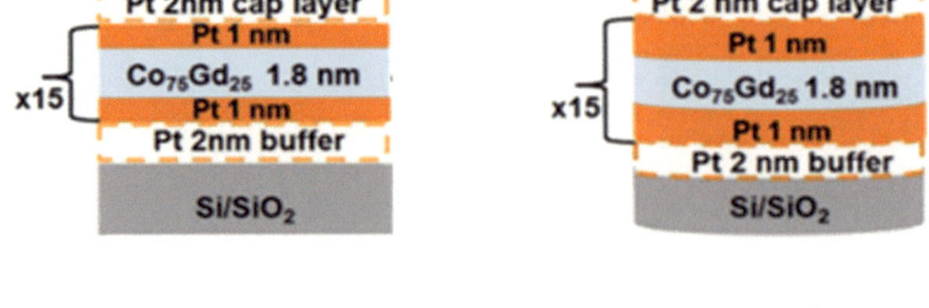

Figure 8.
Representative geometries of the unpatterned (a) and patterned (b) multilayer (shaped as discs) of the CoGd/Pt multilayer. Figure adapted from [27].

sputtering batch. Electron beam lithography (EBL) was firstly performed on the Si/SiO_2 substrate to fabricate the patterned discs. More details about the methods used in the sample fabrication can be found in Ref. [27].

3.1 CoGd layer magnetic orientation

3.1.1 Magnetization behavior

As a first step, the magnetic properties of CoGd/Pt-based multilayer unpatterned sample were characterized. We used a vibrating sample magnetometer (VSM) and a SQUID to measure the hysteresis loops at ambient conditions and the temperature-dependent magnetization, respectively, as shown in **Figure 9**.

Hysteresis loops acquired by VSM at room-temperature exhibit the magnetization reversal for in-plane (IP) and out-of-plane (OOP) magnetic fields (**Figure 9(a)**). One may observe that the unpatterned CoGd multilayer presents a perpendicular magnetic anisotropy (PMA). On the other side, the out-of-plane magnetization has a tail-like structure with low remnant magnetization (blue curve) that reflects on the magnetic domain pattern observed at zero-field as it will be shown later. In **Figure 9(b)**, the temperature-dependent magnetization under zero magnetic field shows a behavior previously observed in ferrimagnetic alloys made of 3d transition metals and 4f rare earth elements. Depending on the composition between the alloy elements, the Co and Gd magnetic moments may align along the same or opposite direction, being ordered ferro- or ferrimagnetically. If the magnetic moments are arranged in an anti-parallel alignment, the compensation temperature corresponds to the magnetic moments of the Co and Gd sublattices compensating each other, leading to a minimization of the resultant magnetization [55]. This behavior is an important signature of the sample ferrimagnetic order. Starting from 2 K and without magnetic field, the sample magnetization decreases slowly as the temperature is increased, reaching a magnetization minimum at 154 K. By further increasing the temperature, the magnetization raises until reaching room-temperature (300 K), before gradually reducing around 350–400 K. This peculiar temperature-dependent behavior indicates that the CoGd/Pt based multilayer has a ferrimagnetic order of the Co and Gd magnetic moments.

3.1.2 X-ray spectroscopy and X-ray magnetic circular dichroism

X-ray absorption spectroscopy (XAS) and X-ray magnetic circular dichroism (XMCD) were performed to extract the magnetic orientation of each Co and Gd

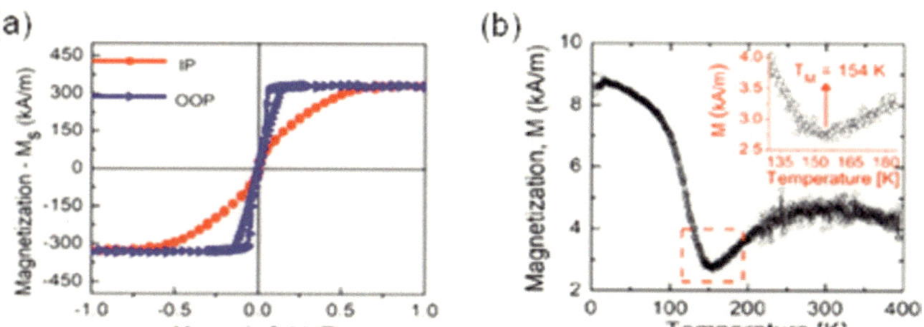

Figure 9.
(a) Hysteresis loops obtained with magnetic field applied in-plane (IP) (red curve) and out-of-plane (OOP) (blue curve). (b) Temperature-dependent magnetization in the absence of magnetic field. Figure adapted from [27].

sublattice by taking advantage of the technique chemical selectivity. It allows us to further confirm the magnetic order of the CoGd alloy at room-temperature. The XAS was obtained under an out-of-plane magnetic field relative to the sample surface and parallel to the incoming X-rays by using a superconducting magnet [56]. Right and left circularly polarized X-rays were used to separate the dependent magnetic orientation of the Co and Gd elements.

Figure 10(a) shows the XAS acquired for Co around the $L_{2,3}$ absorption edges. At the first and more intense peak around 779 eV (L_3 edge), there are two distinct amplitudes: higher (dashed blue line), for left circularly polarized X-rays, and lower (solid red line), for right circularly polarized X-rays. This difference in amplitude is due to the X-ray magnetic circular dichroism that relates the dependence of the element magnetic orientation with the incoming X-ray polarization. Likewise, for Gd, a difference in amplitude is also observed at the most intense peak around 1552 eV at the Gd M_5 edge. However, contrary to the Co case, higher amplitude occurs for right circularly polarized X-rays, while lower absorption for left circularly polarized X-rays (**Figure 10(c)**). These results reflect on the XMCD (difference between the right and left circularly polarized X-ray spectra), as demonstrated in **Figure 10(b, d)**.

Looking at the L_3 and M_5 edges where the XAS maximum occurs, one can see that the XMCD is negative for Co, while it is positive for Gd. This confirms that the magnetic orientation of the Co and Gd are arranged in an antiparallel direction, thus forming a ferrimagnetic CoGd alloy at room-temperature. This result helped us to conduct the magnetic force microscopy images of the CoGd/Pt

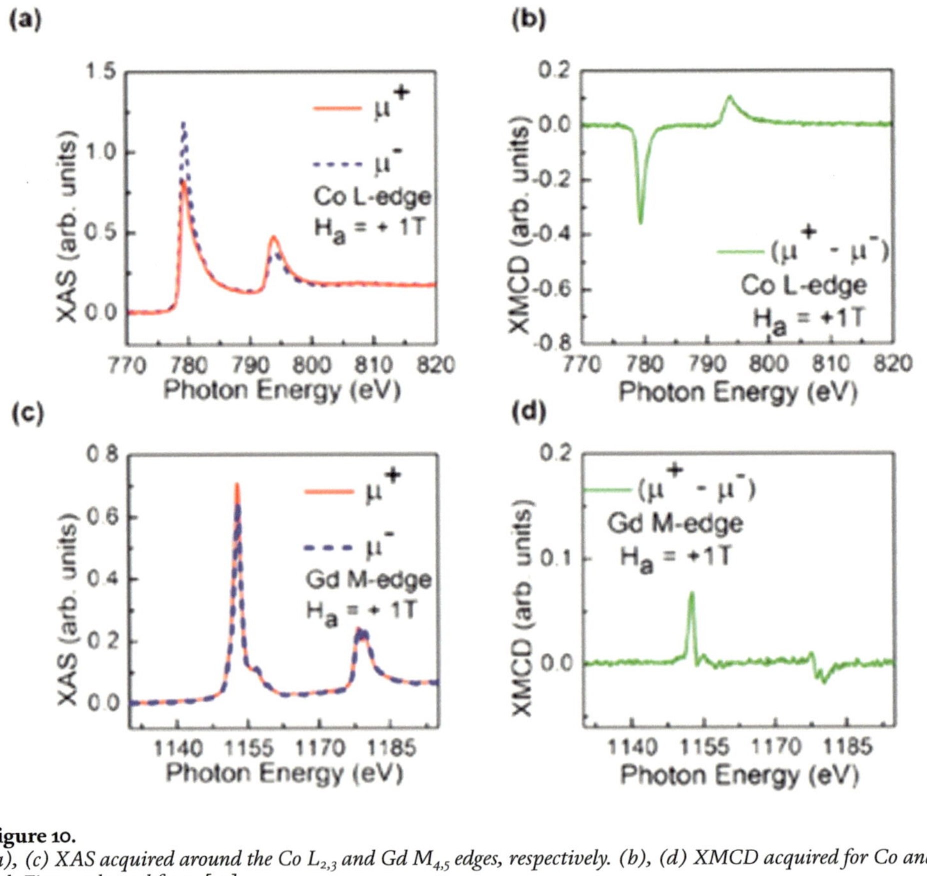

Figure 10.
(a), (c) XAS acquired around the Co $L_{2,3}$ and Gd $M_{4,5}$ edges, respectively. (b), (d) XMCD acquired for Co and Gd. Figure adapted from [27].

multilayer and confirms the magnetic textures observed at ambient conditions as ferrimagnetic domains.

3.2 Magnetic textures

3.2.1 Unpatterned multilayer

Magnetic force microscopy was carried out to image the magnetic domains of the unpatterned sample and observe its features. **Figure 11(a)** shows the image acquired at remanence with alternating red and blue stripes in a typical worm-like configuration. By recalling that the out-of-plane magnetization of the unpatterned sample exhibits low remnant magnetization (see **Figure 9(a)**), it is expected to observe this type of magnetic texture in such CoGd/Pt-based multilayer.

The magnetic domain pattern is characterized by a spatial periodicity of the stripes obtained by using a linear profile on the MFM (dashed line) (**Figure 11(a)**). **Figure 11(b)** exhibits the oscillations along a 1 μm length of the spatial domain distribution of the up and down magnetic stripes, resulting in a wavelength of ~150 nm and domain width Δ of ~90 nm. The MFM image undertaken in the unpatterned multilayer will be used as a guide to evaluate the evolution of the magnetic domains in the confined structures, as will be discussed in the next section.

3.2.2 Confined nanostructures

Here we present the magnetic domain textures of the CoGd/Pt multilayer influenced by the confinement induced by the geometric disc. We produced disc arrays by electron beam lithography and lift-off techniques with different diameters of 1 μm, 650 nm, and 250 nm. **Figure 12** (top) shows the scanning electron microscopy images (SEM) of the arrays.

To visualize better the magnetic domain formation in the nanostructures, individual MFM images of the discs for each diameter are displayed at the bottom of **Figure 12**. **Figure 12** (a, bottom) displays the magnetic domain for the larger disc (1 μm). One may see that the disc supports a magnetic domain pattern like the one observed in the unpatterned multilayer (see **Figure 11(a)**), with alternating up and down magnetic stripes. Therefore, there is no significant influence of the geometric confinement at this specific diameter on the magnetic domain type formation. On the other hand, the geometric confinement leads to a modification of the magnetic stripe periodicity in the disc compared to the unpatterned sample

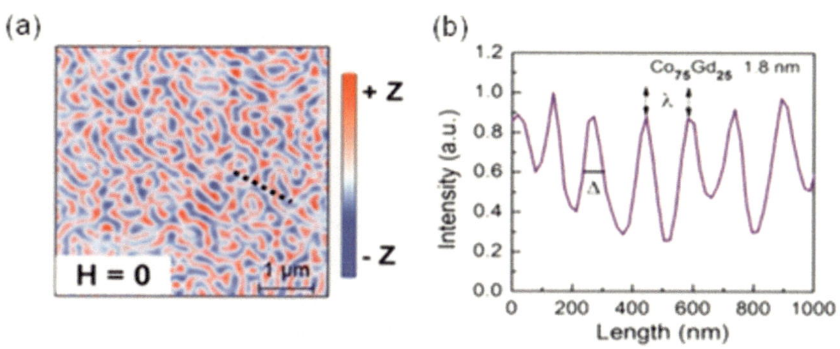

Figure 11.
(a) MFM image of the unpatterned sample exhibiting magnetic stripes (blue and red) with out-of-plane orientation. (b) Domain periodicity extracted by performing line scan on the MFM image. Figure adapted from [27].

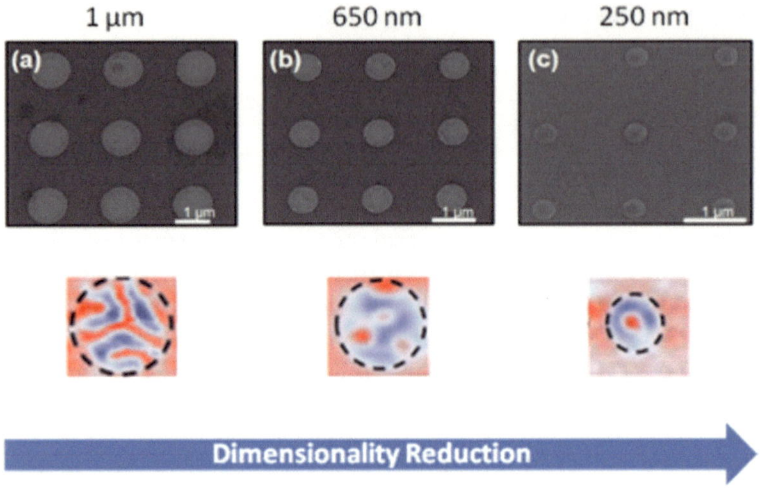

Figure 12.
(Top) Scanning electron microscopy (SEM) images of the disc arrays. (Bottom) Respective MFM images of magnetic domain textures in individual discs. (a) 1 µm, (b) 650 nm, (c) 250 nm. Figure adapted from [27].

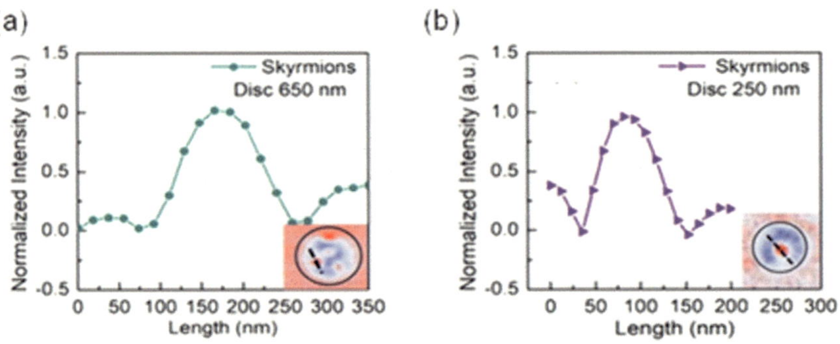

Figure 13.
Profile as a function of the length acquired by using line scan on the multiple skyrmions (a) and single skyrmions (b). Adapted from the supplementary material of reference [27].

(~ 190 nm vs. ~150 nm, respectively). This indicates that the confinement given by the 1 µm diameter increases the distance between up and down magnetic stripes.

The impact of the disc size is more evidenced for the 650 nm diameter disc, where the confinement strongly modifies the magnetic domain pattern. The disc does not support a magnetic stripe domain, but rather the formation of small skyrmions, with a size of ~120 nm (**Figure 13(a)**). It is worth noting that this disc size supports three ferrimagnetic skyrmions at room-temperature and zero magnetic field. By further reducing the disc size to 250 nm, a stable single ferrimagnetic skyrmion is observed in the nanostructure. The single ferrimagnetic skyrmion exhibits a size of ~70 nm (**Figure 13(b)**), which is a very desirable feature for the use of this magnetic structure in future applications [57]. Therefore, by tailoring the disc diameter, we could stabilize magnetic domain textures at room-temperature and zero-field from magnetic stripe configurations, in larger discs, to multiple or single skyrmions in smaller discs. Furthermore, the reduction of the disc size and hence increased confinement modifies not only the magnetic domain type but also allows the existence of multiple or single ferrimagnetic skyrmions. This control can be used in new applications based on the dynamic mode of coupled skyrmions such as neuromorphic computing [58].

4. Conclusion

To summarize, in this chapter we presented two recent approaches to stabilize skyrmions at zero magnetic field and room-temperature. First, ferromagnetic skyrmions were observed in unpatterned symmetric Pd/Co/Pd multilayer by tuning both the remnant magnetization and perpendicular magnetic anisotropy. The PMA enhancement in combination with high remnant magnetization found for the thinnest Co layer (0.2 nm) yields the formation of ferromagnetic skyrmions at ambient conditions without the assistance of a magnetic field. By comparing the experimental 165 nm and simulated skyrmion sizes, we are able to estimate the non-vanishing DMI around 0.4–0.8 mJm^{-2}, as an additional sample parameter required to stabilize skyrmions even in nominally symmetric multilayers. The DMI obtained in the micromagnetic modeling emerges mainly owing to the imperfections (roughness, intermixing) at the Co/Pd interfaces, allowing the formation of zero-field magnetic domains and isolated skyrmions.

Second, ferrimagnetic skyrmions were stabilized in confined geometric discs made of Pt/CoGd-based multilayers. The ferrimagnetic order of the multilayer is confirmed by dependent-temperature SQUID measurement and room-temperature XMCD. By adjusting the disc diameter, magnetic stripes in larger discs evolve to multiple and single ferrimagnetic skyrmions in smaller discs. The single skyrmion observed in the 250 nm diameter exhibits a size of ~70 nm. This sub-100 nm size and quasi antiferromagnetic arrangement of the skyrmions are very desirable both for fundamental physics understanding regarding the skyrmion Hall effect and technological applications based on antiparallel coupled skyrmions. These mechanisms are robust to find skyrmions at ambient conditions and zero-field by engineering symmetric multilayers made of ferro- or ferrimagnetic layers.

Acknowledgements

This work was supported by the Brazilian funding agencies *Fundação de Amparo à Pesquisa do Estado de São Paulo* (FAPESP) (Projects No. 2012/51198-2, 2017/10581-1; 2020/07397-7) and *Conselho Nacional de Desenvolvimento Científico e Tecnológico* (CNPq) (Projects 436573/2018-0 and 436573/2018-0). We thank the Microfabrication Laboratory of the Brazilian Nanotechnology National Laboratory (LNNano, CNPEM) for support in growing sample facilities and Laboratory of Materials and Low Temperatures of the University of Campinas (UNICAMP), for the use of VSM and SQUID techniques. We also thank the Planar Grating Monochromator (PGM) beamline at the Brazilian Synchrotron Light Laboratory (LNLS, CNPEM) for XAS and XMCD measurements.

Conflict of interest

The authors declare no conflict of interest.

Author details

Jeovani Brandão[1], Marcos Vinicius Puydinger dos Santos[2] and Fanny Béron[2*]

1 Brazilian Synchrotron Light Laboratory (LNLS), Brazilian Center for Research in Energy and Materials (CNPEM), Campinas, SP, Brazil

2 Gleb Wataghin Institute of Physics, University of Campinas (UNICAMP), Campinas, SP, Brazil

*Address all correspondence to: fberon@ifi.unicamp.br

References

[1] A. N. Bogdanov and D. A. Yablonskii, Thermodynamically stable "vortices" in magnetically ordered crystals. The mixed state of magnets. Sov. Phys. JETP 68, 101 (1989).

[2] Ryu, KS., Thomas, L., Yang, SH. *et al.* Chiral spin torque at magnetic domain walls. Nature Nanotech 8, 527533 (2013).

[3] Emori, S., Bauer, U., Ahn, S.-M., Martinez, E. & Beach, G. S. D. Current-driven dynamics of chiral ferromagnetic domain walls. Nat. Mater. 12, 611 (2013).

[4] Chen, G. *et al.* Tailoring the chirality of magnetic domain walls by interface engineering. Nat. Commun. 4, 2671 (2013).

[5] Brataas, A. Chiral domain walls move faster. Nat. Nanotech. 8, 485 (2013).

[6] Nagaosa, N. & Tokura, Y. Topological properties and dynamics of magnetic skyrmions. Nat. Nanotechol. 8, 899 (2013).

[7] Shibata, K. *et al.* Towards control of the size and helicity of skyrmions in helimagnetic alloys by spin–orbit coupling. Nature Nanotech. 8, 723 (2013).

[8] A. Hrabec, Z. Luo, L. J. Heyderman, and P. Gambardella, Synthetic chiral magnets promoted by the Dzyaloshinskii-Moriya interaction. Appl. Phys. Lett. 117, 130503 (2020)

[9] Borie, B., Kehlberger, A., Wahrhusen, J., Grimm, H. & Kläui, M. Geometrical dependence of domain-wall propagation and nucleation fields in magnetic-domain-wall sensors. Phys. Rev. Applied 8, 024017 (2017).

[10] J. Brandão, A. Mello, F. Garcia, L.C. Sampaio. Trajectory and chirality of vortex domain walls in ferromagnetic nanowires with an asymmetric branch. J. Appl. Phys., 121, 093905 (2017).

[11] B. Borie, J. Wahrhusen, H. Grimm, M. Kläui. Geometrically enhanced closed-loop multi-turn sensor devices that enable reliable magnetic domain wall motion. Appl. Phys. Lett. 111, 242402 (2017).

[12] Allwood, D. A. *et al.* Magnetic domain wall logic. Science 309, 1688 (2005).

[13] Currivan-Incorvia, J. A. *et al.* Logic circuit prototypes for three-terminal magnetic tunnel junctions with mobile domain walls. Nat. Commun. 7, 10275 (2016).

[14] Luo, Z., Hrabec, A., Dao, T.P. *et al.* Current-driven magnetic domain-wall logic. Nature 579, 214 (2020).

[15] Parkin, S. S. P., Hayashi, M. & Thomas, L. Magnetic domain-wall racetrack memory. Science 320, 190 (2008).

[16] Tomasello, R., Martinez, E., Zivieri, R. *et al.* A strategy for the design of skyrmion racetrack memories. Sci Rep 4, 6784 (2014).

[17] R. D. McMichael, M. J. Donahue. Head to head domain walls structures in thin magnetic stripes. IEEE Trans. Magn. 33, 4167 (1997).

[18] T. Shinjo *et al.*, Magnetic vortex observation in circular dots. Science 289, 930 (2000).

[19] U. K. Rößler, A. N. Bogdanov, C. Pfleiderer, Spontaneous skyrmion ground states in magnetic metals.Nature 442, 797 (2006).

[20] Moreau-Luchaire, C. *et al.* Additive interfacial chiral interaction in multilayers for stabilization of small

individual skyrmions at room temperature. Nat. Nanotech. 11, 444 (2016).

[21] Upadhyaya, P.; Yu, G.; Amiri, P. K.; Wang, K. L. Electric-Field Guiding of Magnetic Skyrmions. Phys. Rev. B: Condens. Matter Mater. Phys. 2015, 92, 134411.

[22] Ogawa, N., Seki, S. & Tokura, Y. Ultrafast optical excitation of magnetic skyrmions. Sci. Rep. 5, 9552 (2015).

[23] Woo, S. *et al.* Observation of room temperature magnetic skyrmions and their current-driven dynamics in ultrathin metallic films. Nat. Mater. 15, 501 (2016).

[24] Bogdanov, A. N. & Rößler, U. K. Chiral symmetry breaking in magnetic thin films and multilayers. Phys. Rev. Lett. 87, 037203 (2001).

[25] Zhang, S. *et al.* Topological computation based on direct magnetic logic communication. Sci. Rep. 5, 15773 (2015).

[26] Senfu Zhang et al. New J. Phys. 17, 023061 (2015).

[27] Brandão, J., Dugato, D. A., Puydinger dos Santos, M. V. & Cezar, J. C. Evolution of zero-field ferrimagnetic domains and skyrmions in exchange-coupled Pt/CoGd/Pt confined nanostructures: implications for antiferromagnetic devices. ACS Appl. NanoMater. 2, 7532 (2019).

[28] Cortés-Ortuño, D., Wang, W., Beg, M. *et al.* Thermal stability and topological protection of skyrmions in nanotracks. Sci. Rep. 7, 4060 (2017).

[29] Fert, A., Cros, V. & Sampaio, J. Skyrmions on the track. Nature Nanotech. 8, 152 (2013).

[30] Meyer, S., Perini, M., von Malottki, S. *et al.* Isolated zero field sub-10 nm skyrmions in ultrathin Co films. Nat Commun. 10, 3823 (2019).

[31] Mühlbauer, S. *et al.* Skyrmion lattice in a chiral magnet. Science 323, 915 (2009).

[32] Yu, X. Z. *et al.* Real-space observation of a two-dimensional skyrmion crystal. Nature 465, 901 (2010).

[33] Yu, X. Z. *et al.* Near room-temperature formation of a skyrmion crystal in thin-films of the helimagnet FeGe. Nat. Mater. 10, 106 (2011).

[34] Dzialoshinskii, I. E. Thermodynamic theory of "weak" ferromagnetism in antiferromagnetic substances. Sov. Phys. JETP 5, 1259(1957).

[35] Soumyanarayanan, A., Raju, M., Gonzalez Oyarce, A. *et al.* Tunable room-temperature magnetic skyrmions in Ir/Fe/Co/Pt multilayers. Nature Mater. 16, 898 (2017).

[36] Fert, A., Reyren, N. & Cros, V. Magnetic skyrmions: advances in physics and potential applications. Nat. Rev. Mater. 2, 17031 (2017).

[37] Hervé, M., Dupé, B., Lopes, R. *et al.* Stabilizing spin spirals and isolated skyrmions at low magnetic field exploiting vanishing magnetic anisotropy. Nat Commun. 9, 1015 (2018).

[38] Romming, N., Kubetzka, A., Hanneken, C., von Bergmann, K. & Wiesendanger, R. Field-dependent size and shape of single magnetic skyrmions. Phys. Rev. Lett. 114, 177203 (2015).

[39] Boulle, O. *et al.* Room-temperature chiral magnetic skyrmions in ultrathin magnetic nanostructures. Nat. Nanotechnol. 11, 449 (2016).

[40] Yu, G. *et al.* Room-temperature skyrmions in an antiferromagnet-based

heterostructure. Nano Lett. 18, 980 (2018).

[41] Zhang, S. *et al.* Direct writing of room temperature and zero field skyrmion lattices by a scanning local magnetic field. Appl. Phys. Lett. 112, 132405 (2018).

[42] Guang, Y., Bykova, I., Liu, Y. *et al.* Creating zero-field skyrmions in exchange-biased multilayers through X-ray illumination. Nat Commun. 11, 949 (2020).

[43] Yao Guang, Yong Peng, Zhengren Yan, Yizhou Liu, Junwei Zhang, Xue Zeng, Senfu Zhang, Shilei Zhang, David M. Burn, Nicolas Jaouen, Jinwu Wei, Hongjun Xu, Jiafeng Feng, Chi Fang, Gerrit Laan, Thorsten Hesjedal, Baoshan Cui, Xixiang Zhang, Guoqiang Yu, Xiufeng Han. Electron Beam Lithography of Magnetic Skyrmions. Advanced Materials, 32, 2003003 (2020).

[44] Pollard, S. D. *et al.* Observation of stable Néel skyrmions in Co/Pd multilayers with Lorentz transmission electron microscopy. Nat. Commun. 8, 14761-14761 (2017).

[45] A. Hrabec, N. A. Porter, A. Wells, M. J. Benitez, G. Burnell, S. McVitie, D. McGrouther, T. A. Moore, and C. H. Marrows, Phys. Rev. B 90, 020402 (2014).

[46] A. V. Davydenko, A. G. Kozlov, A. G. Kolesnikov, M. E. Stebliy, G. S. Suslin, Y. E. Vekovshinin, A. V. Sadovnikov, and S. A. Nikitov, Dzyaloshinskii-Moriya interaction in symmetric epitaxial Co/Pd (111)xN superlattices with different numbers of Co/Pd bilayers, Phys. Rev. B 99, 014433 (2019).

[47] Brandão, J., Dugato, D.A., Seeger, R.L. *et al.* Observation of magnetic skyrmions in unpatterned symmetric multilayers at room temperature and zero magnetic field. Sci. Rep. 9, 4144 (2019).

[48] Liu, Z. *et al.* Thickness dependent magnetization dynamics of perpendicular anisotropy Co/Pd multilayer films. J. Magn. Magn. Mater. 323, 1623 (2011).

[49] Behera, A. K. *et al.* Size and shape of skyrmions for variable Dzyaloshinskii Moriya interaction and uniaxial anisotropy. J. Phys. D: Appl. Phys. 51, 285001 (2018).

[50] Jiang, W. *et al.* Direct observation of the skyrmion Hall effect. Nature Phys. 13, 162 (2016).

[51] Zhang, X. *et al.* Antiferromagnetic Skyrmion: Stability, Creation and Manipulation. Sci. Rep. 6, 24795 (2016).

[52] Haiyan Xia *et al.* Control and manipulation of antiferromagnetic skyrmions in racetrack. J. Phys. D: Appl. Phys. 50, 505005 (2017).

[53] Buttner, F., Lemesh, I. Beach, G. S. D. Theory of isolated magnetic skyrmions: From fundamentals to room temperature applications. Sci. Rep. 8, 4464 (2018).

[54] Woo, S. *et al.* Current-driven dynamics and inhibition of the skyrmion Hall effect of ferrimagnetic skyrmions in GdFeCo films. Nat. Commun. 9,959 (2018).

[55] Caretta, L; *et al.* Fast current-driven domain walls and small skyrmions in a compensated ferrimagnet. Nat. Nanotechnol. 13, 1154-1160 (2018).

[56] Cezar, J. C.; *et al.* The U11 PGM beam line at the Brazilian National Synchrotron Light Laboratory. J. Phys.: Conf. Ser. 425, 072015 (2013).

[57] Navau, C.; Del-Valle, N.; Sanchez, A. Analytical trajectories of skyrmions in confined geometries: Skyrmionic

racetracks and nanooscillators. Phys. Rev. B: Condens. Matter Mater. Phys. 94, 184104 (2016).

[58] Torrejon, J.; Riou, M.; Araujo, F. A.; Tsunegi, S.; Khalsa, G.; Querlioz, D.; Bortolotti, P.; Cros, V.; Yakushiji, K.; Fukushima, A.; Kubota, H.; Yuasa, S.; Stiles, M. D.; Grollier, J. Neuromorphic computing with nanoscale spintronic oscillators. Nature, 547, 428 (2017).

Chapter 4

Modeling the Bulk and Nanometric Dielectric Functions of Au and Ag

Brahim Ait Hammou, Abdelhamid El Kaaouachi, Abdellatif El Oujdi, Adil Echchelh, Said Dlimi, Chi-Te Liang and Jamal Hemine

Abstract

In this work, we model the dielectric functions of gold (Au) and silver (Ag) which are typically used in photonics and plasmonics. The modeling has been performed on Au and Ag in bulk and in nanometric states. The dielectric function is presented as a complex number with a real part and an imaginary part. First, we will model the experimental measurements of the dielectric constant as a function of the pulsation ω by appropriate mathematical functions in an explicit way. In the second part we will highlight the contributions to the dielectric constant value due to intraband and interband electronic transitions. In the last part of this work we model the dielectric constant of these metals in the nanometric state using several complex theoretical models such as the Drude Lorentz theory, the Drude two-point critical model, and the Drude three-point critical model. We shall comment on which model fits the experimental dielectric function best over a range of pulsation.

Keywords: Modeling bulk and nanometric dielectric function, noble metals Au and Ag, interband transitions, intraband transitions, UV and IR pulsations

1. Introduction

All the intrinsic effects corresponding to the process of light-matter interaction are contained in the dielectric function $\varepsilon(\omega)$. In the case of an isotropic material, the optical response is described by following equation

$$\varepsilon(\omega) = \varepsilon_1(\omega) + i\varepsilon_2(\omega) \tag{1}$$

where $\varepsilon(\omega)$ is generally a complex scalar value which depends upon the pulsation ω of the field. If the medium has an anisotropy, this magnitude is in the form of a tensor. It is often convenient to describe the optical response in an equivalent way from the complex refractive index $\tilde{n} = n + i\kappa$ as n denotes the refractive index describing the phase speed of the wave and κ denotes the extinction index describing the absorption of the wave during propagation in the material. These two indices are directly related to the dielectric constant of the material. In fact, the real and imaginary parts of the dielectric function are deduced from the relation:

$$\varepsilon_1 = n^2 - \kappa^2 \tag{2}$$

$$\varepsilon_2 = 2n\kappa \tag{3}$$

Several important physical quantities can be deduced from the complex refractive index $\tilde{n}$ and dielectric function $\varepsilon(\omega)$ such as the reflectivity coefficient R and the attenuation coefficient α. In the fields of photonics, and plasmonics, researchers use the dielectric function in their calculations and investigations [1–8].

2. Modeling of the bulk experimental dielectric function

Here we try to model the experimental dielectric function $\boldsymbol{\varepsilon^{exp}(\omega)}$ of noble metals (Ag and Au) on a wide pulse interval ω with adequate mathematical functions. We consider the measured values of the dielectric function reported by Dold and Mecke [9], Winsemius *et al.* [10], Leveque *et al.* [11], and Thèye *et al.* [12] respectively. These measurements are summarized by Rakiç *et al.* [13] in **Figure 1** (for Ag), and in **Figure 2** (for Au).

2.1 Modeling the experimental dielectric function of bulk Ag

In this part, we will model the real and imaginary parts of the experimental dielectric function of Ag (**Figure 1**). For this we will divide the values of the pulsation ω into several intervals in order to allow us to determine the best fit to suitable mathematical functions over a certain interval. All the results will be presented in **Tables 1–8**.

2.1.1 Modeling the real part of the bulk dielectric function of Silver $\varepsilon^{exp}_{R\_Ag}(\omega)$

For $1,899 \times 10^{14}\ rad/s \leq \omega \leq 1,275 \times 10^{15}\ rad/s$

$$\varepsilon^{exp}_{R\_Ag}(\omega) = \varepsilon^0_{R_0} + A_1\left(1 - e^{-\omega/t_1}\right) + A_2\left(1 - e^{-\omega/t_2}\right) \tag{4}$$

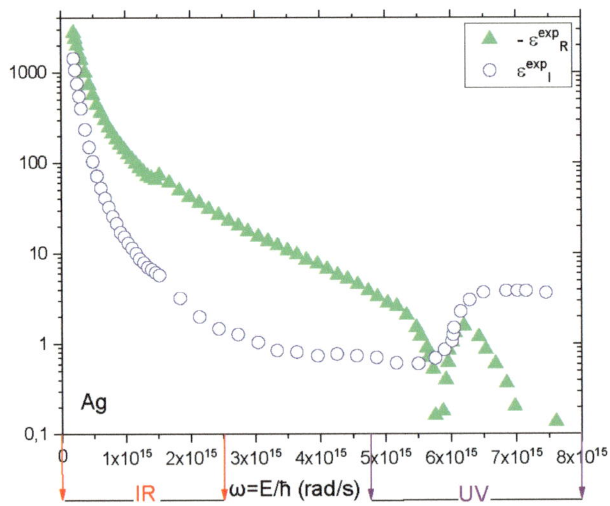

Figure 1.
Real part (▲) and imaginary part (○) of the dielectric function of bulk Ag ([adapted] with permission from [Ref. [13]])© The Optical Society.

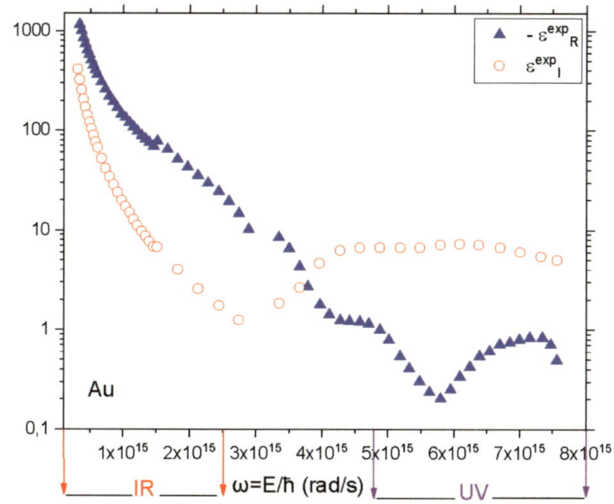

Figure 2.
Real part(▲) and imaginary part (○) of the dielectric function of bulk Au([adapted] with permission from [Ref. [13]])© The Optical Society.

Parameter	$\varepsilon^0_{R_0}$	A_1	t_1(rad/s)	A_2	t_2(rad/s)	R-Square (COD)
Value	10499.713	−1300.876	3.813×10^{14}	−9163.320	1.217×10^{14}	0.9999

Table 1.
Values of the model parameters.

Parameter	$\varepsilon^1_{R_0}$	A'_1	t'_1(rad/s)	A'_2	t'_2(rad/s)	A_3	t_3(rad/s)	R-Square (COD)
Value	0.305	83.392	1.091×10^{15}	86.023	1.091×10^{15}	86.639	1.091×10^{15}	0,99101

Table 2.
Values of the model parameters.

Parameter	$\varepsilon^2_{R_0}$	A''_1	t''_1 (rad/s)	A''_2	t''_2 (rad/s)	R-Square (COD)
Value	5.281	−0.006	-9.588×10^{14}	−0.006	-9.576×10^{14}	0.99671

Table 3.
Values of the model parameters.

Parameter	$\varepsilon^3_{R_0}$	A	ω_C (rad/s)	B (rad/s)	R-Square (COD)
Value	0.868	0.735	-9.816×10^{14}	3.483×10^{14}	0.96592

Table 4.
Values of the model parameters.

Parameter	$\varepsilon^4_{R_0}$	$\omega_{C'}$ (rad/s)	H	ω_1 (rad/s)	ω_2 (rad/s)	R-Square (COD)
Value	2.539	7.140×10^{15}	−2.399	6.978×10^{14}	3.783×10^{24}	0.99613

Table 5.
Values of the model parameters.

Parameter	$\varepsilon_{I_0}^0$	B_1	ω_0 (rad/s)	V_1 (rad/s)	R-Square (COD)
Value	−23218.730	585.784	2.267×10^{14}	6.910×10^{13}	0.99998
	B_2	V_2 (rad/s)	B_3	V_3 (rad/s)	
	306.100	1.844×10^{14}	23235.380	2.514×10^{18}	

Table 6.
Values of the model parameters.

Parameter	$\varepsilon_{I_0}^1$	B'	TD (rad/s)	τ (rad/s)	R-Square (COD)
Value	7.076	−6.327	1.38101×10^{15}	4.92416×10^{14}	0.9983

Table 7.
Values of the model parameters.

Parameter	B_1'	B_2'	p	$\mathrm{Ln}\omega_{01}$ (rad/s)	h_1 (rad/s)$^{-1}$	$\mathrm{Ln}\omega_{02}$ (rad/s)
Value	1.278	1.633	−8.811	9.897×10^{15}	$1.262 \times 10–^{16}$	6.159×10^{15}
	h_2 (rad/s)$^{-1}$	R-Square (COD)				
	$3.945 \times 10–^{15}$	0.99733				

Table 8.
Values of the model parameters.

For $1,275 \times 10^{15}\ rad/s \le \omega \le 4,873 \times 10^{15}\ rad/s$

$$\varepsilon_{R\_Ag}^{exp}(\omega) = A_1' e^{-\omega/t_1'} + A_2' e^{-\omega/t_2'} + A_3 e^{-\omega/t_3} + \varepsilon_{R_0}^1 \tag{5}$$

For $4.873 \times 10^{15}\ rad/s \le \omega \le 5.650 \times 10^{15}\ rad/s$

$$\varepsilon_{R\_Ag}^{exp}(\omega) = A_1'' e^{-\omega/t_1''} + A_2'' e^{-\omega/t_2''} + \varepsilon_{R_0}^2 \tag{6}$$

For $5.650 \times 10^{15}\ rad/s \le \omega \le 6.198 \times 10^{15}\ rad/s$

$$\varepsilon_{R\_Ag}^{exp}(\omega) = \varepsilon_{R_0}^3 + A \sin\left(\pi \frac{\omega - \omega_C}{B}\right) \tag{7}$$

For $6.198 \times 10^{15}\ rad/s \le \omega \le 7.605 \times 10^{15}\ rad/s$

$$\begin{cases} \varepsilon_{R\_Ag}^{exp}(\omega) = \varepsilon_{R_0}^4 + He^{-\frac{1}{2}\left(\frac{\omega-\omega_{C'}}{\omega_1}\right)^2} for(\omega < \omega_{C'}) \\ \varepsilon_{R\_Ag}^{exp}(\omega) = \varepsilon_{R_0}^4 + He^{-\frac{1}{2}\left(\frac{\omega-\omega_{C'}}{\omega_2}\right)^2} for(\omega \ge \omega_{C'}) \end{cases} \tag{8}$$

2.1.2 Modeling the imaginary part of the bulk dielectric function of Silver $\varepsilon_{I\_Ag}^{exp}(\omega)$

For $1,900 \times 10^{14} rad/s \le \omega \le 1,340 \times 10^{15} rad/s$

$$\varepsilon_{I\_Ag}^{exp}(\omega) = \varepsilon_{I_0}^0 + B_1 e^{-(\omega-\omega_0)/V_1} + B_2 e^{-(\omega-\omega_0)/V_2} + B_3 e^{-(\omega-\omega_0)/V_3} \tag{9}$$

For $1.340 \times 10^{15} rad/s \le \omega \le 4.251 \times 10^{15} rad/s$

$$\begin{cases} \varepsilon_{I\_Ag}^{exp}(\omega) = \varepsilon_{I_0}^{1} for(\omega < TD) \\ \varepsilon_{I\_Ag}^{exp}(\omega) = \varepsilon_{I_0}^{1} + B'\left(1 - e^{-\frac{(\omega - TD)}{\tau}}\right) for(\omega \geq TD) \end{cases} \quad (10)$$

For $4.252 \times 10^{15} rad/s \leq \omega \leq 7.453 \times 10^{15} rad/s$

$$\varepsilon_{I\_Ag}^{exp}(\omega) = B_1' + (B_2' - B_1')\left[\frac{p}{1 + 10^{(Ln\omega_{01} - \omega)h_1}} + \frac{1 - p}{1 + 10^{(Ln\omega_{02} - \omega)h_2}}\right] \quad (11)$$

2.2 Modeling the experimental dielectric function of bulk Au

In this section, we will model the real and imaginary parts of the experimental dielectric function of bulk Au (**Figure 2**). For this we will proceed in the same way as for bulk Ag by dividing the values of the pulsation ω into various intervals. All the results are listed in **Tables 9–15**.

2.2.1 Modeling the real part ofthe bulk dielectric function of Gold $\varepsilon_{R\_Au}^{exp}(\omega)$

For $3.319 \times 10^{14} rad/s \leq \omega \leq 1.515 \times 10^{15} rad/s$

$$\varepsilon_{R\_Au}^{exp}(\omega) = a_0 + a_1\omega + a_2\omega^2 + a_3\omega^3 + a_4\omega^4 + a_5\omega^5 + a_6\omega^6 + a_7\omega^7 + a_8\omega^8 + a_9\omega^9 \quad (12)$$

For $1.515 \times 10^{15} rad/s \leq \omega \leq 3.345 \times 10^{15} rad/s$

$$\varepsilon_{R\_Au}^{exp}(\omega) = a_1' + (a_2' - a_1')\left[\frac{q}{1 + 10^{(Ln\omega'_{01} - \omega)g_1}} + \frac{1 - q}{1 + 10^{(Ln\omega'_{02} - \omega)g_2}}\right] \quad (13)$$

For $3.345 \times 10^{15} rad/s \leq \omega \leq 4.271 \times 10^{15} rad/s$

$$\varepsilon_{R\_Au}^{exp}(\omega) = b_0 + b_1\omega + b_2\omega^2 + b_3\omega^3 + b_4\omega^4 + b_5\omega^5 \quad (14)$$

For $4.271 \times 10^{15} rad/s \leq \omega \leq 7.560 \times 10^{15} rad/s$

$$\varepsilon_{R\_Au}^{exp}(\omega) = c_0 + c_1\omega + c_2\omega^2 + c_3\omega^3 + c_4\omega^4 + c_5\omega^5 + c_6\omega^6 + c_7\omega^7 + c_8\omega^8 + c_9\omega^9 \quad (15)$$

2.2.2 Modeling the imaginary part ofthe bulk dielectric function of Gold $\varepsilon_{I\_Au}^{exp}(\omega)$

For $3.047 \times 10^{14} rad/s \leq \omega \leq 1.511 \times 10^{15} rad/s$

$$\varepsilon_{I\_Au}^{exp}(\omega) = d_0 + d_1\omega + d_2\omega^2 + d_3\omega^3 + d_4\omega^4 + d_5\omega^5 + d_6\omega^6 + d_7\omega^7 + d_8\omega^8 + d_9\omega^9 \quad (16)$$

For $1.511 \times 10^{15} rad/s \leq \omega \leq 4.265 \times 10^{15} rad/s$

$$\varepsilon_{R\_Au}^{exp}(\omega) = f_0 + f_1\omega + f_2\omega^2 + f_3\omega^3 + f_4\omega^4 + f_5\omega^5 \quad (17)$$

Parameter	a_0	a_1 **(rad/s)**$^{-1}$	a_2**(rad/s)**$^{-2}$	a_3 **(rad/s)**$^{-3}$	a_4 **(rad/s)**$^{-4}$	**R-Square (COD)**
Value	**4835.830**	**-8.633×10^{-12}**	**-6.930×10^{-26}**	**3.779×10^{-40}**	**-8.746×10^{-55}**	0.9999
	a_5 (rad/s)$^{-5}$	a_6 (rad/s)$^{-6}$	a_7 (rad/s)$^{-7}$	a_8 (rad/s)$^{-8}$	a_9 (rad/s)$^{-9}$	
	1.170×10^{-69}	-9.674×10^{-85}	4.893×10^{-100}	-1.392×10^{-115}	1.710×10^{-131}	

Table 9.
Values of the model parameters.

Parameter	a'_1	a'_2	q	$\mathrm{Ln}\omega'_{01}$ **(rad/s)**	g_1 **(rad/s)**$^{-1}$	$\mathrm{Ln}\omega'_{02}$ **(rad/s)**	g_2 **(rad/s)**$^{-1}$	**R-Square (COD)**
Value	7.786	138.106	0.896	1.495×10^{15}	-1.388×10^{-15}	2.641×10^{15}	-3.366×10^{-15}	0.9996

Table 10.
Values of the model parameters.

Parameter	b_0	b_1 **(rad/s)**$^{-1}$	b_2 **(rad/s)**$^{-2}$	b_3 **(rad/s)**$^{-3}$	b_4 **(rad/s)**$^{-4}$	b_5 **(rad/s)**$^{-5}$	**R-Square (COD)**
Value	-50227.482	6.432×10^{-11}	-3.281×10^{-26}	8.338×10^{-42}	-1.056×10^{-57}	5.337×10^{-74}	0.9997

Table 11.
Values of the model parameters.

Parameter	c_0	c_1 **(rad/s)**$^{-1}$	c_2 **(rad/s)**$^{-2}$	c_3 **(rad/s)**$^{-3}$	c_4 **(rad/s)**$^{-4}$	**R-Square (COD)**
Value	**271340.234**	**-4.224×10^{-10}**	**2.912×10^{-25}**	**-1.167×10^{-40}**	**2.995×10^{-56}**	0.99872
	c_5 (rad/s)$^{-5}$	c_6 (rad/s)$^{-6}$	c_7 (rad/s)$^{-7}$	c_8 (rad/s)$^{-8}$	c_9 (rad/s)$^{-9}$	
	-5.112×10^{-72}	5.801×10^{-88}	-4.220×10^{-104}	1.786×10^{-120}	-3.351×10^{-137}	

Table 12.
Values of the model parameters.

Parameter	d_0	d_1 (rad/s)$^{-1}$	d_2 (rad/s)$^{-2}$	d_3 (rad/s)$^{-3}$	d_4 (rad/s)$^{-4}$	R-Square (COD)
Value	9873.897	-9.845×10^{-11}	4.479×10^{-25}	-1.188×10^{-39}	2.002×10^{-54}	0.9999
	d_5 (rad/s)$^{-5}$	d_6 (rad/s)$^{-6}$	d_7 (rad/s)$^{-7}$	d_8 (rad/s)$^{-8}$	d_9 (rad/s)$^{-9}$	
	-2.212×10^{-69}	1.598×10^{-84}	-7.276×10^{-100}	1.895×10^{-115}	-2.152×10^{-131}	

Table 13.
Values of the model parameters.

Parameter	f_0	f_1 (rad/s)$^{-1}$	f_2 (rad/s)$^{-2}$	f_3 (rad/s)$^{-3}$	f_4 (rad/s)$^{-4}$	f_5 (rad/s)$^{-5}$	R-Square (COD)
Value	115.516	-1.891×10^{-13}	1.299×10^{-28}	-4.538×10^{-44}	7.892×10^{-60}	-5.361×10^{-76}	0.9973

Table 14.
Values of the model parameters.

Parameter	j_0	j_1 (rad/s)$^{-1}$	j_2 (rad/s)$^{-2}$	j_3 (rad/s)$^{-3}$	j_4 (rad/s)$^{-4}$	R-Square (COD)
Value	−53432.059	1.046×10^{-13}	-8.722×10^{-26}	4.106×10^{-41}	-1.210×10^{-56}	0.9974
	j_5 (rad/s)$^{-5}$	j_6 (rad/s)$^{-6}$	j_7 (rad/s)$^{-7}$	j_8 (rad/s)$^{-8}$	j_9 (rad/s)$^{-9}$	
	2.324×10^{-72}	-2.918×10^{-88}	2.316×10^{-104}	-1.056×10^{-120}	2.111×10^{-137}	

Table 15.
Values of the model parameters.

For $4.265 \times 10^{15}\ rad/s \leq \omega \leq 7.560 \times 10^{15}\ rad/s$

$$\varepsilon_{I\_Au}^{exp}(\omega) = j_0 + j_1\omega + j_2\omega^2 + j_3\omega^3 + j_4\omega^4 + j_5\omega^5 + j_6\omega^6 + j_7\omega^7 + j_8\omega^8 + j_9\omega^9 \quad (18)$$

All the results of these different models for both bulk Ag and bulk Au are plotted in the following **Figures 3** and **4**.

In both cases, the experimental results can be well fitted by the models of experimental dielectric function in both its real and imaginary parts with mathematical functions with high accuracy.

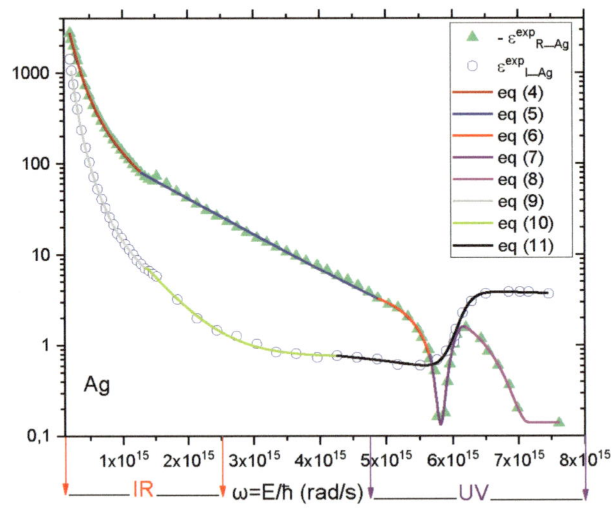

Figure 3.
Real (▲) and imaginary (○) parts of the experimental dielectric function of bulkAg (Ref. [13]). Solid colored curves represent, by pulse intervals, the different mathematical models used in the modeling.

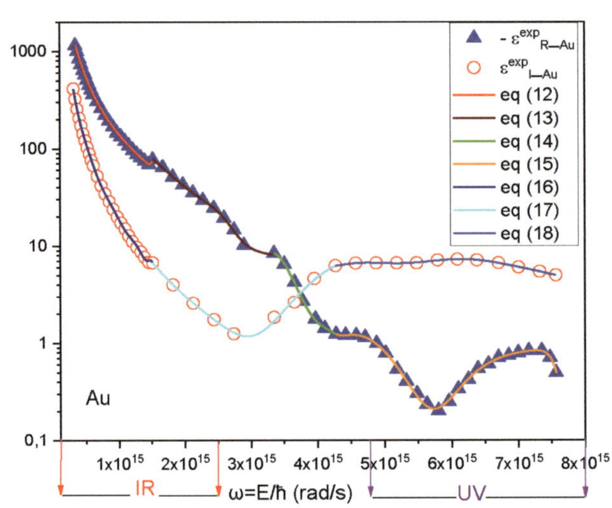

Figure 4.
Real (▲) and imaginary (○) parts of the experimental dielectric function of bulkAu (Ref. [13]). Solid colored curves represent, by pulse intervals, the different mathematical models used in the modeling.

3. Highlighting the contribution of interband and intraband transitions in the expression of dielectric function

In metals, there aretwo types of contribution in the dielectric function, namely contribution of interband transitions denoted $\varepsilon^{iB}(\omega)$ and that of the intraband transitions denoted $\varepsilon^{D}(\omega)$. The dielectric function can be written as the sum of two terms

$$\varepsilon(\omega) = \varepsilon^{D}(\omega) + \varepsilon^{iB}(\omega) \tag{19}$$

The first term corresponds to the intraband component of the dielectric function. It is referred to the optical transitions of a free electron from the conduction band to a higher energy level of the same band. The second term corresponds to the interband component of the dielectric constant. It is referred to optical transitions between the valence bands (mainly d band and s-p conduction bands). Due to Pauli's exclusion principle, an electron from a valence band can only be excited to the conduction band. There is therefore an energy threshold E^{IB} for interband transitions which is placed in the visible band for Au and near the UV region for Ag. This component is often overlooked in the infrared range (this is valid only for alkali metals) where the optical response is dominated by intraband absorption. This type of transitions dominates the optical response beyond E^{IB}.

The intraband part $\varepsilon^{D}(\omega)$ of the dielectric function is described by the well-known free-electronorDrude Model [14]:

$$\varepsilon^{D}(\omega) = 1 - \frac{\omega_p^2}{\omega(\omega + i\gamma)} \tag{20}$$

where γ is the collision rate (probability of collision per unit of time). εω = ε1ω + iε2ω.

We note that $\gamma = \tau^{-1}$ and τ is the elastic diffusion time

$$\tau = \frac{l_0}{v_f} \tag{21}$$

where l_0 is the mean free path of electrons and v_f is the Fermi speed.

In the Drude model, there appears a pulsation called the plasmon frequency of a bulk metal given by

$$\omega_p = \sqrt{\frac{n_c e^2}{\varepsilon_0 m_{eff}}} \tag{22}$$

Where ε_0 is dielectric constant in vacuum.

The electronic structure of bulk noble metals such as Ag and Au, the respective values of the conduction electron density n_c, the effective mass m_{eff}, the Fermi speed v_f conduction electrons and the mean free path of electron l_0, are listed in the following **Table 16** [15]:

3.1 Contribution of intraband transitions to dielectric function

The intraband dielectric function described by the Drude model [14] as denoted $\varepsilon^{D}(\omega)$ can be written as

Metal	$n_c\ (\times 10^{28} m^{-3})$	$\frac{m_{eff}}{m_e}$	$v_f(\text{nm/fs})$	$l_0(\text{nm})$
Ag	5.86	0.96	1.39	55
Au	5.90	0.99	1.40	42

From Phys. Review B.6. 4376(1972).

Table 16.
Electronic properties of Ag and Au.

$$\varepsilon^D(\omega) = \varepsilon_R^D(\omega) + i\varepsilon_I^D(\omega) \tag{23}$$

The real and imaginary parts of the relative dielectric function (intraband) are written as follows

$$\varepsilon_R^D(\omega) = 1 - \frac{\omega_p^2}{\omega^2 + \gamma^2} \tag{24}$$

$$\varepsilon_I^D(\omega) = \frac{\omega_p^2 \gamma}{\omega(\omega^2 + \gamma^2)} \tag{25}$$

Usually, for noble metals $\omega \gg \gamma$ in the near UVrange and up to the near IR, we can write

$$\varepsilon_R^D(\omega) = 1 - \frac{\omega_p^2}{\omega^2} \tag{26}$$

$$\varepsilon_I^D(\omega) = \frac{\omega_p^2 \gamma}{\omega^3} \tag{27}$$

The following **Table 17** shows the values of plasma frequencies and the collision rate of noble metals (Gold, Silver):

The results of the calculations of the contribution of intraband effects to dielectric function are represented in their real and imaginary parts for bulk Ag (**Figure 5**) and for bulk Au (**Figure 6**).

For the noble metals (Gold, Silver), we observed that the real and imaginary parts decrease withincreasing pulsation. The further away from the pulsations corresponding to IR radiation and the closer we get to the pulsations corresponding to UV radiation, these values decrease.

3.2 Contribution of interband transitions to dielectric function

The interband dielectric function denoted $\varepsilon^{iB}(\omega)$ is described by the following term

$$\varepsilon^{iB}(\omega) = \varepsilon_R^{iB}(\omega) + i\varepsilon_I^{iB}(\omega) \tag{28}$$

Metal	ω_p^2 (rad/s)^2	γ (s^{-1})
Ag	1.9428×10^{32}	2.5273×10^{13}
Au	1.8968×10^{32}	3.3333×10^{13}

Table 17.
Plasma frequency and collision rate values for Ag and Au.

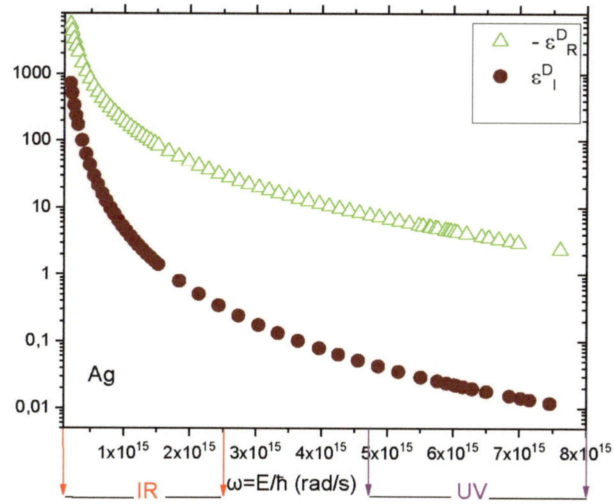

Figure 5.
Real (△) imaginary part (•) of the intraband dielectric function of bulk Ag.

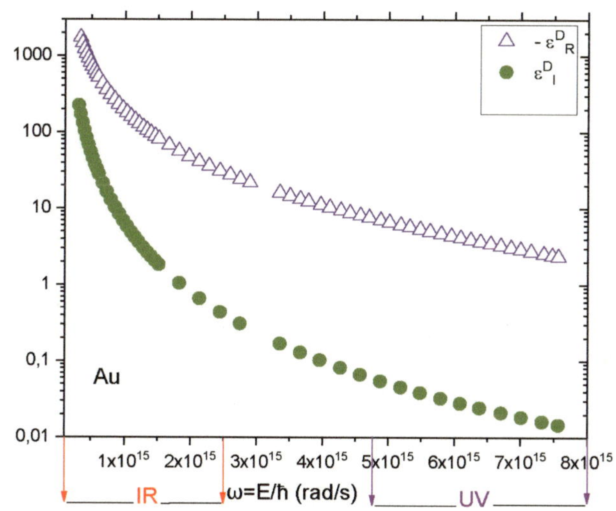

Figure 6.
Real (△) imaginary part (•) of intraband dielectric function of bulk Au.

The real and imaginary parts of the interband dielectric function are written respectively as follows:

$$\varepsilon_{R}^{iB}(\omega) = \varepsilon_{R}^{exp}(\omega) - \varepsilon_{R}^{D}(\omega) \tag{29}$$

$$\varepsilon_{I}^{iB}(\omega) = \varepsilon_{I}^{exp}(\omega) - \varepsilon_{I}^{D}(\omega) \tag{30}$$

where

$$\varepsilon^{exp}(\omega) = \varepsilon_{R}^{exp}(\omega) + i\varepsilon_{I}^{exp}(\omega) \tag{31}$$

Here $\varepsilon^{exp}(\omega)$, $\varepsilon_{R}^{exp}(\omega)$, *and* $\varepsilon_{I}^{exp}(\omega)$ are the experimental values, real and imaginary parts of the complex dielectric function, respectively.

The results of the calculations of the contribution of interband transitions to real and imaginary parts of dielectric function are represented respectively for bulk Ag (**Figure 7**) and for bulk Au (**Figure 8**).

As shown in **Figure 7**, the real part of the contribution of interband effects to the dielectric function of bulk Ag decreases with increasing pulsation in the IR radiation domain then is still almost constant with small variations from $\omega = 1.5 \times 10^{15}$rad/s and this until the end of the UV radiation range. Concerning the imaginary part of this contribution, it is almost less important than the real part. It decreases with increasing pulsation in the range corresponding to IR radiation, then varies very little to the value of the pulsation $\omega = 6 \times 10^{15}$rad/s in the range of UV radiation, in this area increase from 0.4 to 1.4 and finally remains constant for the rest of the UV pulses. The imaginary part becomes superior to the real part for UV pulses higher than the value $\omega = 6 \times 10^{15}$rad/s.

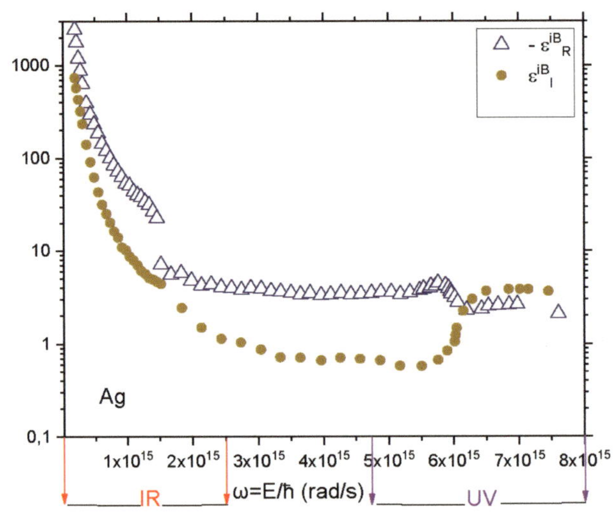

Figure 7.
Real (△) and imaginary part (●) of the interband dielectric function of bulk Ag.

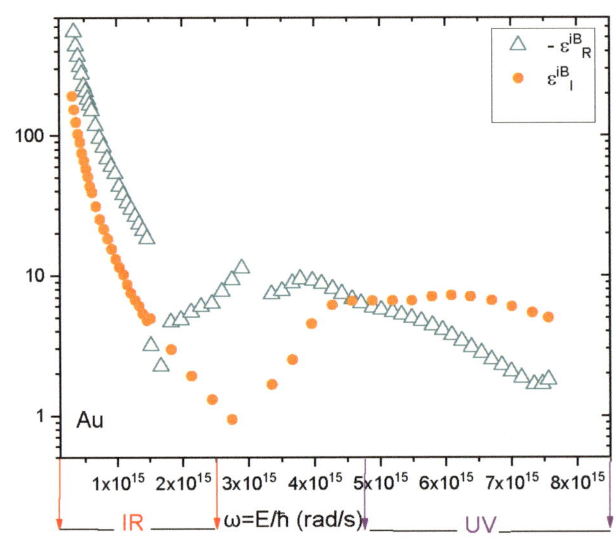

Figure 8.
Real (△) and imaginary part (●) of the interband dielectric function of bulk Au.

We note that for Gold, the real and imaginary parts of the contribution of interband transitions decrease by increasing the values of the pulses in the IR domain to the value $\omega = 4.5 \times 10^{15}$rad/s with a real part almost higher than the imaginary part. For pulsation values in the UV range above the value $\omega = 4.5 \times 10^{15}$rad/s, with a slight variation the imaginary part becomes superior to the real part.

In **Figures 9** and **10** (for Ag) and **Figures 11** and **12** (for Au), we have presented experimental values, the contributions of intraband and interband transitions to the real and imaginary parts of the dielectric function.

Concerning the real part, we note that, the real part due to the intraband transitions noted $\varepsilon_R^D(\omega)$ is in very good agreement with the experimental values in the full domain corresponding to IR radiation and up to the value $\omega = 4 \times 10^{15}$rad/s. In this range of pulsations, we can conclude that the participation in the dielectric function

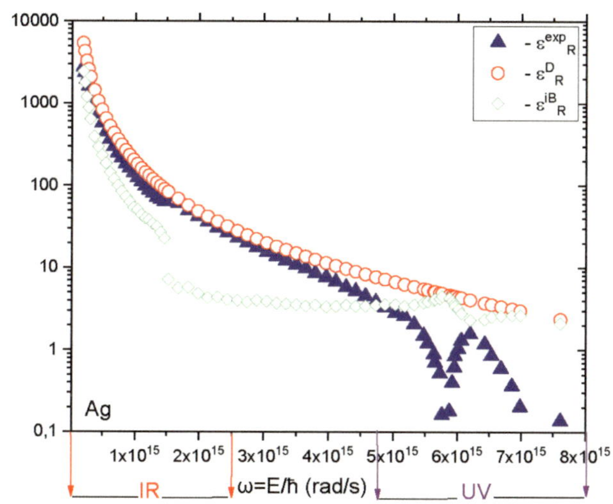

Figure 9.
Real parts of the dielectric function of Ag: experimental values (▲) ([Ref. [13]), intraband transitions (○), interband transitions (◇).

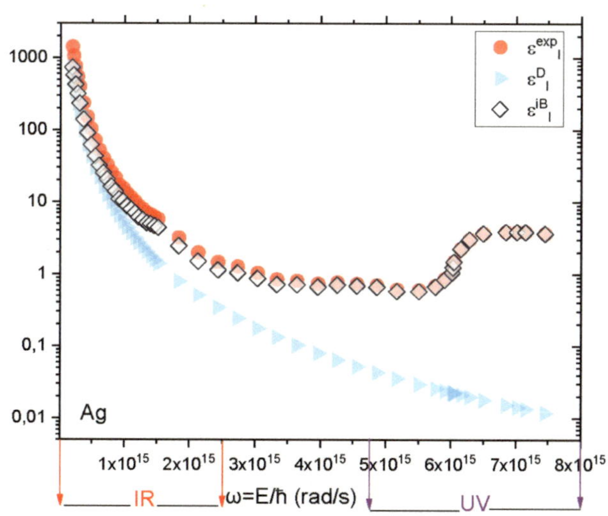

Figure 10.
Imaginary parts of the dielectric function of Ag: experimental values (●) ([Ref. [13]), intraband transitions (▶), interband transitions (◇).

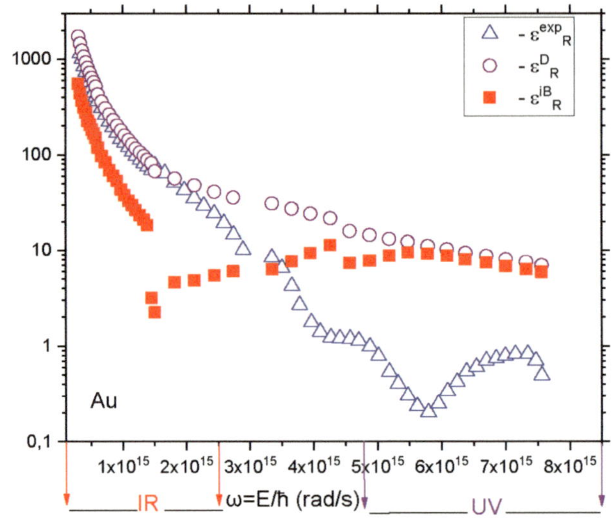

Figure 11.
Real parts of the dielectric function of Au: experimental values (△) ([Ref. [13]), intraband transitions (○), interband transitions (■).

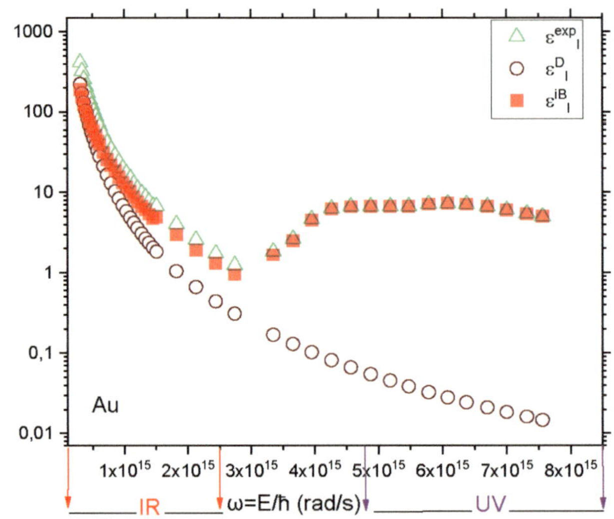

Figure 12.
Imaginary parts of the dielectric function of Au: experimental values (△) ([Ref. [13]), intraband transitions (○),interband transitions (■).

due to inter-band transitions is negligible compared to that of intra-band transitions. For $\omega > 4.5 \times 10^{15} rad/s$ in the range of UV radiation, the two contributions due to intraband and interband transitions are equivalent but differ from the experimental measurements.

Concerning the imaginary part, we note that the imaginary part due to the interband transitions noted $\varepsilon_I^{iB}(\omega)$ is practically confused with the imaginary experimental part in the whole field of pulsations from the IR domain to the end of the UV domain. This indicates that the imaginary part due to intraband transitions is negligible compared to the imaginary part due to interband transitions.

For Gold, we note that, the real part due to the intraband transitions noted $\varepsilon_R^D(\omega)$ is in very good agreement with the real experimental values in the whole range corresponding to IR radiation and up to the value $\omega = 2 \times 10^{15}$rad/s. In this range of pulsations, we can conclude that the participation in the dielectric function due to interband transitions is negligible compared to that of intraband transitions. For $\omega > 2 \times 10^{15}$rad/s and up to the UV radiation domain, the two contributions due to intraband and interband transitions differ significantly from the experimental measurements. These two contributions are very similar in the field of UV for $\omega \geq 2.5 \times 10^{15}$rad/s.

Concerning the imaginary part, as for Silver, we see that the imaginary part due to the interband transitions noted $\varepsilon_I^{iB}(\omega)$ is practically confused with the imaginary experimental part in the whole field of pulsations from the IR domain to the end of the UV domain. This indicates that the imaginary part due to intraband transitions is negligible compared to the imaginary part due to interband transitions.

4. Modeling the dielectric functions of nanometri cAg and nanometric Au

In this section, we study the dielectric functions of nanometric Ag and nanometric Au. They are composed of a few tens to several thousand atoms. Their very small characteristic dimensions, in the nanometer range (i.e. well under optical wavelengths), give rise to extraordinary electronic and optical properties that cannot be observed in bulk materials. These properties are clearly influenced by the size, form of the nanoparticle and the nature of the host environment. We consider the measured values of dielectric function used in the previous paragraphs, and try to model those using theoretical models for nanometals such as the Drude Lorentz (DL) model, the Drude two-point critical model DCP and the Drude three-point critical model DCP3.

The optical properties of metallic nanoparticles are dominated by the collective oscillation of conduction electrons induced by interaction with electromagnetic radiation (IR, UV).

The collective excitation of nanoparticle conduction electrons gives them new optical properties; we consider the following two effects:

- Plasmons guided along a metallic film of nanometric cross-section (1D confinement).
- Surface plasmons located in a metallic particle of nanometric size (0D confinement).

4.1 Drude Lorentz (DL) model

For the study of resonant nanostructures, it is important to have a good description of the permittivity of the metal in a large frequency band. For this purpose, the validity band of the Drude Model is often extended by adding Lorentzian terms [16] depending in the following form

$$\varepsilon_{DL}(\omega) = \varepsilon_\infty - \frac{\omega_D^2}{\omega^2 + i\gamma\omega} + \sum_{l=1}^{2} \frac{f_l \Omega_l^2}{\Omega_l^2 - \omega^2 - i\Gamma_l \omega} \tag{32}$$

The dielectric function described by the Drude Lorentz model is written as follows

$$\varepsilon_{DL}(\omega) = \varepsilon_{DL}^R(\omega) + i\varepsilon_{DL}^{im}(\omega) \tag{33}$$

where:
The real part of the dielectric function according to the DL model

$$\varepsilon^{R}_{DL}(\omega) = \varepsilon_\infty - \frac{\omega_D{}^2}{\omega^2+\gamma^2} + \frac{f_1\Omega_1^2(\Omega_1^2-\omega^2)}{\left(\Omega_1^2-\omega^2\right)^2+(\Gamma_1\omega)^2} + \frac{f_2\Omega_2^2(\Omega_2^2-\omega^2)}{\left(\Omega_2^2-\omega^2\right)^2+(\Gamma_2\omega)^2} \tag{34}$$

The imaginary part of the dielectric function according to the DL model:

$$\varepsilon^{im}_{DL}(\omega) = \frac{\gamma\omega_D{}^2}{\omega^3+\gamma^2\omega} + \frac{f_1\Omega_1^2\Gamma_1\omega}{\left(\Omega_1^2-\omega^2\right)^2+(\Gamma_1\omega)^2} + \frac{f_2\Omega_2^2\Gamma_2\omega}{\left(\Omega_2^2-\omega^2\right)^2+(\Gamma_2\omega)^2} \tag{35}$$

The studies of Vial and Laroche [16] on the permittivity of Auand Ag metals used in their model with the parameters are listed in **Table 18.**

The results of modeling the experimental dielectric function in its real and imaginary parts using the DL model are shown in **Figures 13** and **14** for Au(in **Figures 15** and **16** for Ag).

4.2 Drude model with two critical points DCP

In order to describe the metal in the largest possible range of pulsations, another formula describing the two-point critical Drude model (DCP) [16] will appear in this paragraph.

The dielectric function of Au and Ag is can be expressed from as [17]:

$$\varepsilon_{DCP}(\omega) = \varepsilon_\infty - \frac{\omega_D^2}{\omega^2+i\gamma\omega} + \sum_{l=1}^{2} A_l\Omega_l\left[\frac{e^{i\phi}}{\Omega_l-\omega-i\Gamma_l} + \frac{e^{-i\phi}}{\Omega_l+\omega+i\Gamma_l}\right] \tag{36}$$

The dielectric function described by the Drude two-critical-point model is written as follows

$$\varepsilon_{DCP}(\omega) = \varepsilon^{R}_{DCP}(\omega) + i\varepsilon^{im}_{DCP}(\omega) \tag{37}$$

where
The real part of the dielectric function according to the DCP model:

	ε_∞	ω_D (rad/s)	γ (rad/s)	f_1	
Au	**6.2137**	**1.3323×10^{16}**	**1.3235×10^{14}**	**3.4620**	
Ag	**1.7984**	**1.3359×10^{16}**	**8.7167×10^{13}**	**3.0079**	
	Ω_1 (rad/s)	Γ_1 (rad/s)	f_2	Ω_2 (rad/s)	Γ_2 (rad/s)
Au	4.7914×10^{15}	2.1367×10^{15}	−3.4886	4.2111×10^{14}	4.5572×10^{17}
Ag	8.1635×10^{15}	4.3785×10^{17}	2.3410	3.8316×10^{16}	6.0574×10^{16}

From J. Phys. D. Appl. Phys. ***40****,7154 (2007).*

Table 18.
Optimized parameters of the Drude Lorentz model for noble metals (Au and Ag).

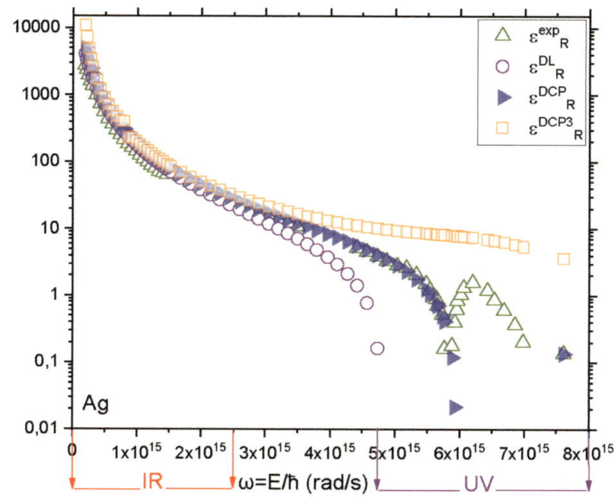

Figure 13.
Real part of the dielectric function of nanometric Ag: experimental values (△) ([Ref. [13]), the DL model (○), the DCP model (▶) and the DCP3model (□).

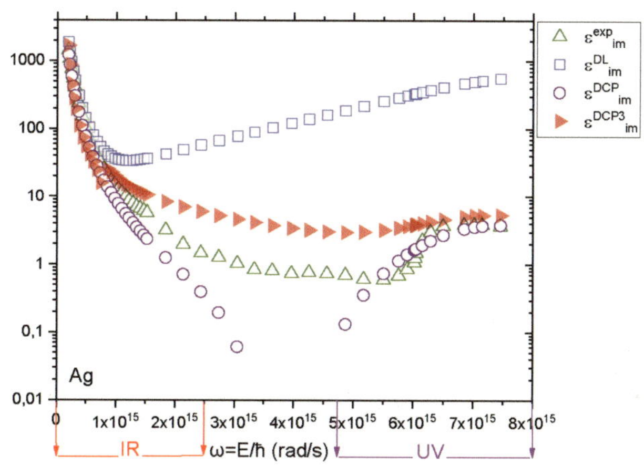

Figure 14.
Imaginary part of the dielectric function of nanometric Ag:experimental values (△) ([Ref. [13]), the DL model (□), the DCPmodel (○), and the DCP3 model (▶).

$$\varepsilon_{DCP}^{R}(\omega) = \varepsilon_{\infty} - \frac{{\omega_D}^2}{\omega^2 + \gamma^2} + \sum_{l=1}^{2} A_l \Omega_l \left[\frac{(\Omega_l - \omega)\cos\phi_l - \Gamma_l \sin\phi_l}{(\Omega_l - \omega)^2 + \Gamma_l^2} + \frac{(\Omega_l + \omega)\cos\phi_l - \Gamma_l \sin\phi_l}{(\Omega_l + \omega)^2 + {\Gamma_l}^2} \right] \tag{38}$$

The imaginary part of the dielectric function according to the DCP model:

$$\varepsilon_{DCP}^{im}(\omega) = \frac{\gamma {\omega_D}^2}{\omega^3 + \gamma^2 \omega} + \sum_{l=1}^{2} A_l \Omega_l \left[\frac{(\Omega_l - \omega)\sin\phi_l + \Gamma_l \cos\phi_l}{(\Omega_l - \omega)^2 + \Gamma_l^2} + \frac{(\Omega_l + \omega)\sin\phi_l - \Gamma_l \cos\phi_l}{(\Omega_l + \omega)^2 + \Gamma_l^2} \right] \tag{39}$$

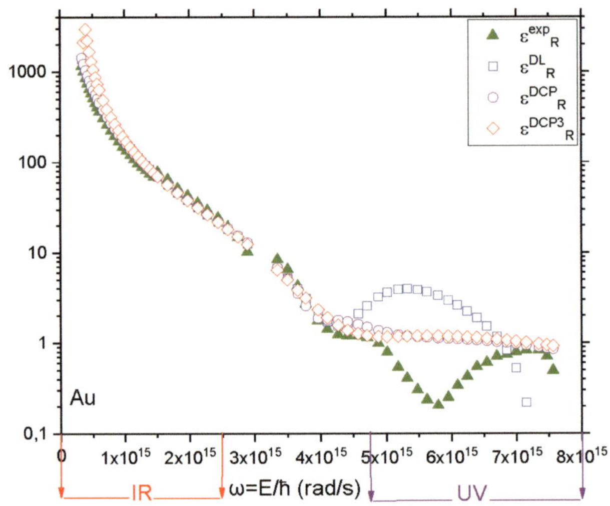

Figure 15.
Real part of the dielectric function of nanometric Au: experimental values (▲) ([Ref. [13]), the DLmodel (□), the DCP model (○), and the DCP3 model (◇).

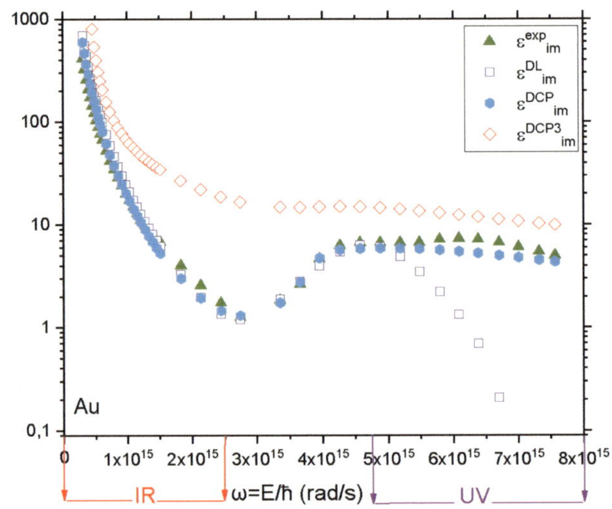

Figure 16.
Imaginary part of dielectric of nanometric Au: experimental values (▲) ([Ref. [13]), the DL model (□), the DCP model (●), and theDCP3 model (◇).

The work of Alexandre Vial's [17] on the permittivity of the noble metals (Gold, Silver) made model with the parameters are listed in **Table 19**.

The results of modeling the experimental dielectric function in its real and imaginary parts using the Drude two-point DCP critical point model are shown in **Figures 13** and **14** for Au (in **Figures 15** and **16** for Ag).

4.3 Drude model with three critical points DCP3

The DCP3 model describes the response of the dielectric function in a wider pulsation band, it should be noted that the DCP3 model gives a very good description of the dielectric function of noble metals; it is expressed by the relation [18]:

	ε_∞	ω_D (rad/s)	γ (rad/s)	A_1	ϕ_1 (rad)	Ω_1 (rad/s)
Au	1.1431	1.3202×10^{16}	1.0805×10^{14}	0.26698	−1.2371	3.8711×10^{15}
Ag	15.833	1.3861×10^{16}	4.5841×10^{13}	1.0171	−0.93935	6.6327×10^{15}
	A_2	ϕ_2 (rad)	Ω_2 (rad/s)	Γ_2 (rad/s)	Γ_1 (rad/s)	
Au	3.0834	−1.0968	4.1684×10^{15}	2.3555×10^{15}	4.4642×10^{14}	
Ag	15.797	1.8087	9.2726×10^{17}	2.3716×10^{17}	1.6666×10^{15}	

From Appl. Phys.B 93, 140 (2008).

Table 19.
Optimized parameters of the Drude two-point critical point model DCP, dielectric function for noble metals (Gold, Silver).

$$\varepsilon_{DCP3}(\omega) = \varepsilon_\infty + \frac{\sigma/\varepsilon_0}{i\omega} + \sum_{l=1}^{3} A_l\Omega_l\left[\frac{e^{i\phi}}{\Omega_l - \omega - i\Gamma_l} + \frac{e^{-i\phi}}{\Omega_l + \omega + i\Gamma_l}\right] \tag{40}$$

The dielectric function described by the Drude two-critical-point model is written as follows:

$$\varepsilon_{DCP3}(\omega) = \varepsilon^{R}_{DCP3}(\omega) + i\varepsilon^{im}_{DCP3}(\omega) \tag{41}$$

Where:
The real part of the dielectric function in the DCP3 model:

$$\varepsilon^{R}_{DCP3}(\omega) = \varepsilon_\infty + \sum_{l=1}^{3} A_l\Omega_l\left[\frac{(\Omega_l - \omega)\cos\phi_l - \Gamma_l\sin\phi_l}{(\Omega_l - \omega)^2 + \Gamma_l^2} + \frac{(\Omega_l + \omega)\cos\phi_l - \Gamma_l\sin\phi_l}{(\Omega_l + \omega)^2 + \Gamma_l{}^2}\right] \tag{42}$$

The imaginary part of the dielectric function in the DCP3 model:

$$\varepsilon^{im}_{DCP3}(\omega) = \frac{-\sigma/\varepsilon_0}{\omega} + \sum_{l=1}^{3} A_l\Omega_l\left[\frac{(\Omega_l - \omega)\sin\phi_l + \Gamma_l\cos\phi_l}{(\Omega_l - \omega)^2 + \Gamma_l{}^2} - \frac{(\Omega_l + \omega)\sin\phi_l + \Omega_l\cos\phi_l}{(\Omega_l + \omega)^2 + \Gamma_l{}^2}\right] \tag{43}$$

The parameters of this model are given in **Table 20**.

The results of modeling the real and imaginary parts of experimental dielectric function using the Drude three-point critical point model DCP3 are shown in **Figures 13** and **14** for Au (in **Figures 15** and **16** for Ag).

Concerning the real part of the dielectric function of nanometric Ag; the model that is very much in agreement with the experiment up to the value of the pulsation $\omega \approx 6 \times 10^{15}$ rad/s, placed in the UV region, is the DCP model. The other models are also valid up to the value of $\omega \approx 3 \times 10^{15}$ rad/s. This interval covers the whole pulsation zone in the IR domain.

For the imaginary part of nanometric Ag, we find that for pulsations located in the IR domain and less than $\omega < 1.5 \times 10^{15}$rad/s, the most appropriate model with experience in this range is the DCP3 model. For pulsation values 1.5×10^{15}rad/s $< \omega < 6 \times 10^{15}$rad/s, the most suitable model to the measured values of the

	ε_∞	σ/ε_0	A_1	ϕ_1 (rad)	Ω_1 (rad/s)	Γ_1 (rad/s)	A_2
Au	1.1156	27.825	0.5548	2.8463	4.506×10^{16}	5.09×10^{16}	679.7606
Ag	1.4783	8.7191	1.007	−0.9621	6.617×10^{15}	1.7415×10^{15}	5377.4512
	ϕ_2 (rad)	Ω_2 (rad/s)	Γ_2 (rad/s)	A_3	ϕ_3 (rad)	Ω_3 (rad/s)	Γ_3 (rad/s)
Au	−0.0998	3.4587×10^{14}	3.064E13	3.5244	4.6586	3.5832×10^{15}	1.68784×10^{15}
Ag	−0.0092	1.3545×10^{14}	6.56505×10^{12}	2.6077	−2.8539	8.1007×10^{14}	8.7193×10^{12}

From Superlattices and Microstructures ***47****, 67 (2009).*

Table 20.
Optimized parameters of Drude at three critical points DCP3 of the dielectric function of noble metals (Au and Ag).

dielectric function is still the DCP3 model. For $\omega > 6 \times 10^{15}$rad/s in the UV range, the two models DCP and DCP3 are very close to the experiment. The DCP3 model is better than the DCP model.

In the case of the real part of the dielectric function of nanometric Au, the DL, DCP, and DCP3 models are allin very good agreement with the experiment up to the value of the pulsation $\omega \approx 5 \times 10^{15}$rad/s; which is the beginning of the UV radiation region. Beyond this value, the three models deviate from the measured values (**Figure 15**). From the value of $\omega \approx 6.75 \times 10^{15}$rad/s two models DCP and DCP3 agree well the experimental values.As shown in **Figure 16**, concerning the imaginary part of the dielectric function of nanometricAu, we note that the DCP model is in excellent agreement with the experiment over the whole domain of pulsation values including values corresponding to both IR and UV radiation. The DL model is also close to the experimental values up to the value $\omega \approx 5 \times 10^{15}$rad/s.

5. Conclusion

In this work, we modeled the dielectric function of noble metals (silver and gold) in their bulk and nanometric states. Initially, we modeled the measured dielectric functions of these two metals using explicit mathematical functions and the results are in very good agreement with the experiment. Moreover, we have decomposed these measured values of the dielectric functions; in their real and imaginary parts; into several intervals according to the pulsations that sweep the domains corresponding to IR and UV radiation via the intermediate values. The obtained results are very conclusive, and depending on the pulsation domain studied, it is possible to use the corresponding mathematical function in simulations and calculations. Then, we highlighted the importance of the contributions of intraband and interband transitions in dielectric function for both Ag and Au. For Ag, we note that the imaginary part of the dielectric function due to interband transitions denoted $\varepsilon_I^{iB}(\omega)$ is almost the same with the imaginary experimental part in the whole field of pulsations from the IR domain to the end of the UV domain. This indicates that the imaginary part due to intraband transitions is negligible compared to the imaginary part due to interband transitions. Concerning the real part of the dielectric function, we note that, the real part due to the intraband transitions noted $\varepsilon_R^D(\omega)$ is in very good agreement with the real experimental values in the entire range corresponding to IR radiation and up to the value $\omega = 4 \times 10^{15}$rad/s. In this range of pulsation, we can conclude that the contribution tothe dielectric function due to interband transitions is negligible compared to that of intraband transitions.

In the case of Au, we note that the real part of the dielectric function due for $\omega > 4.5 \times 10^{15}$rad/$s$ and in the field of UV radiation, the two contributions due to intraband and interband transitions are equivalent but they differ from the experiment. to the intraband transitions denoted as $\varepsilon_R^D(\omega)$ is in very good agreement with the real experimental values over the whole range corresponding to IR radiation and up to the value $\omega = 2.5 \times 10^{15}$rad/s. In this range of pulsation, we can conclude that the participation in the dielectric function due to interband transitions is negligible compared to that of intraband transitions. For $\omega > 2.5 \times 10^{15}$rad/$s$ and up to the UV radiation domain, the two contributions due to intraband and interband transitions differ significantly from the experimental results. The two contributions are very similar in the field of UV for $\omega \geq 2.5 \times 10^{15}$rad/s. Concerning the imaginary part, and as for Ag, we note that the imaginary part due to the interband transitions noted $\varepsilon_I^{iB}(\omega)$ is practically the same with the imaginary experimental part in the whole field of pulsations from the IR domain to the end of the UV domain. This indicates that the imaginary part due to intraband transitions is negligible compared to the imaginary part due to interband transitions.

In the last part of this paper, we have modeled the dielectric functions of Ag and Au, using theoretical models that deal with nanometric systems such as the Drude Lorenz model, the Drude two-point critical model, and the Drude three-point critical model.In the case of nanometic Ag, the real part of the dielectric function model agrees well with the experiment up to the value of the pulsation $\omega \approx 6 \times 10^{15}$rad/s, which is located in the UV radiation region, is the DCP model. The other models are also valid up to the value of $\omega \approx 3 \times 10^{15}$rad/s. This interval covers the entire pulsation zone located in the IR domain. For the imaginary part of Ag, we find that for pulsation located in the IR domain and less than $\omega < 1.5x10^{15}$rad/s, the most appropriate model is the DCP3 model. For pulsation values 1.5×10^{15}rad/s $< \omega < 6 \times 10^{15}$rad/s, model that deviates the least from the measured values of the dielectric function is still the DCP3 model. For $\omega > 6 \times 10^{15}$rad/s in the UV domain, the two models DCP and DCP3 are very close to the experience with a better approach using the DCP3 model.

For nanometric Au, concerning the real part of the dielectric function, the three models DL, DCP, and DCP3 are all in very good agreement with the experiment up to the value of the pulsation $\omega \approx 5 \times 10^{15}$rad/s; which is the beginning of the UV radiation region. Beyond this value, the three models deviate from the experiment. From the value of $\omega \approx 6.75 \times 10^{15}$rad/s the two models DCP and DCP3 meet the experimental values. Concerning the imaginary part, we note that the DCP model is in very good agreement with the experiment on the whole range of pulsation values including values corresponding to both IR and UV radiation. The DL model is also very close to the experimental values up to the value $\omega \approx 5 \times 10^{15}$ rad/s.

Acknowledgements

We are grateful to Professor Uwe Thumm who hosted us for three months in his James R. Macdonald laboratory at the Kansas State University in the USA, and who offered us an opportunity to collaborate on this subject, as part of the Fulbright Grant Merit Award.

Author details

Brahim Ait Hammou[1], Abdelhamid El Kaaouachi[1*], Abdellatif El Oujdi[2], Adil Echchelh[2], Said Dlimi[3], Chi-Te Liang[4] and Jamal Hemine[2]

1 Faculty of Sciences of Agadir, Materials and Physicochemistry of the Atmosphere and Climate Group, BP, Agadir, Morocco

2 Faculty of Sciences Ibn Tofail, Laboratory of Energetic Engineering and Materials, Kenitra, Morocco

3 Faculty of Sciences of Agadir, Physics Department, BP, Agadir, Morocco

4 Department of Physics, National Taiwan University, Taipei, Taiwan

*Address all correspondence to: kaaouachi21@yahoo.fr

References

[1] M. J. Ambrosio and U. Thumm, "Energy-resolved attosecond interferometric photoemission from Ag(111) and Au(111) surfaces," Phys. Rev. A **97**, 043431 (2018).

[2] M. J. Ambrosio and U. Thumm, "Electronic structure effects in spatiotemporally resolved photoemission interferograms of copper surfaces," Phys. Rev. A **96**, 051403(R) (2017).

[3] S. R. Leone, C. W. McCurdy, J. Burgdörfer, L. S. Cederbaum, Z. Chang, N. Dudovich, J. Feist, C. H. Greene, M. Ivanov, R. Kienberger, U. Keller, M. F. Kling, Z.-H. Loh, T. Pfeifer, A. N. Pfeiffer, R. Santra, K. Schafer, A. Stolow, U. Thumm, and M. J. J. Vrakking,"What will it take to observe processes in 'real time'?," Nat. Photon. **8**, 162-166 (2014).

[4] F.Calegari, G. Sansone, S. Stagira, C. Vozzi, and M. Nisoli,"Advances in attosecond science," J. Phys. B **49**, 062001 (2016).

[5] Yanzeng Li, Margaret Kocherga, Serang Park, Marc Lata, Micheal McLamb, Glenn Boreman, Thomas A. Schmedake, and Tino Hofmann, "Optical dielectric function of Si(2,6-bis (benzimidazol-2′-yl)pyridine)$_2$ determined by spectroscopic ellipsometry," Opt. Mater. Express 9, 3469–3475 (2019).

[6] M. J. Ambrosio and U. Thumm, "Comparative time-resolved photoemission from the Cu(100) and Cu(111)," surfaces Phys. Rev. A 94, 063424 (2016).

[7] N. Manrique and H. Riascos, "Estimation of Dielectric Constant and Thickness of Copper Thin Films Using Surface Plasmon Resonance," in *Latin America Optics and Photonics Conference*, OSA Technical Digest (Optical Society of America, 2018), paper Th4A.16.

[8] F. Roth, C. Lupulescu, E. Darlatt, A. Gottwald, and W. Eberhardt,"Angle resolved Photoemission from Cu single crystals; Known Facts and a few Surprises about the Photoemission Process," J. Electron Spectrosc. **208**, 2-10 (2016).

[9] B. Dold and R. Mecke, "Optische Eigenschaften von Edelmetallen, Übergangsmetallen und deren Legierungen im Infrarot (1. Teil)," Optik**22,** 435–446 (1965).

[10] P. Winsemius, H. P. Langkeek, and F. F. van Kampen,"Structure dependence of the optical properties of Cu, Ag and Au," Physica**79B**, 529–546 (1975).

[11] G. Leveque, C. G. Olson, and D. W. Lynch, "Reflectance spectra and dielectric functions of Ag in the region of interband transitions," Phys. Rev. B **27,** 4654–4660 (1983).

[12] M. L. Thèye, "Investigation of the optical properties of Au by means of thin semitransparent films," Phys. Rev. B **2**, 3060-3078 (1970).

[13] Aleksandar D. Rakić, Aleksandra B. Djurišić, Jovan M. Elazar, and Marian L. Majewski,"Optical properties of metallic films for vertical-cavity optoelectronic devices," Applied Optics **37**, 5271-5284 (1998).

[14] C. Kittel, "Introduction to solid states," Wiley, New York, (1971).

[15] P. B. Johnson and R.W. Christy,"Optical constants of the nobles metals," Phys. Rev. B**6**, 4370-4379 (1972).

[16] A.Vial and T. Laroche,"Description of dispersion proprieties of metals by means of the critical points model and application to the study of resonant structures using the FDTD method,"

J. Phys. D. Appl. Phys. **40**, 7152-7158 (2007).

[17] A. Vial and T. Laroche, "Comparison of gold and silver dispersion laws suitable for FDTD simulations," Appl. Phys.B **93**, 139-149 (2008).

[18] J.Y. Lu and Y.H. Chang, "Implementation of an efficient dielectric function into the finite difference time domain method for simulating the coupling localized surface plasmons of nanostructures," Superlattices and Microstructures **47**, 60-65 (2009).

Chapter 5

Magnetic Properties of Heusler Alloys and Nanoferrites

Devinder Singh and Kuldeep Chand Verma

Abstract

In this chapter, results of our recent investigations on the structural, microstructural and magnetic properties of Cu-based Heusler alloys and MFe_2O_4 (M = Mn, Fe, Co, Ni, Cu, Zn) nanostructures will be discussed. The chapter is divided into two parts, the first part describes growth and different characterizations of Heusler alloys while in the second part magnetic properties of nano-ferrites are discussed. The $Cu_{50}Mn_{25}Al_{25-x}Ga_x$ (x = 0, 2, 4, 8 and 10 at %) alloys have been synthesized in the form of ribbons. The alloys with x ≤ 8 show the formation of Heusler single phase of the Cu_2MnAl structure. Further increase of Ga content gives rise to the formation of γ-Cu_9Al_4 type phase together with Cu_2MnAl Heusler phase. The alloys are ferromagnetically ordered and the saturation magnetization (M_s) decreases slightly with increasing Ga concentration. Annealing of the ribbons significantly changes the magnetic properties of $Cu_{50}Mn_{25}Al_{25-x}Ga_x$ alloys. The splitting in the zero field cooled (ZFC) and field cooled (FC) magnetization curves at low temperature has been observed for alloys. Another important class of material is Nanoferrites. The structural and magnetization behaviour of spinel MFe_2O_4 nanoferrites are quite different from that of bulk ferrites. X-ray diffraction study revealed spinel structure of MFe_2O_4 nanoparticles. The observed ferromagnetic behaviour of MFe_2O_4 depends on the nanostructural shape as well as ferrite inversion degree. The magnetic interactions in Ce doped $CoFe_2O_4$ are antiferromagnetic that was confirmed by zero field/field cooling measurements at 100 Oe. Log R (Ω) response measurement of $MgFe_2O_4$ thin film was taken for 10–90% relative humidity (% RH) change at 300 K.

Keywords: Heusler alloy, nanoferrites, magnetization, microstructure, sensor

1. Introduction

Heusler alloys were discovered in 1903 when Heusler reported that the addition of sp. elements (Al, In, Sb, Sn or Bi) turn Cu-Mn alloy into a ferromagnetic material even though the alloy contains none of the ferromagnetic elements [1, 2]. Heusler reported that these alloys have stoichiometric composition X_2YZ and exhibit an ordered $L2_1$ crystal structure with space group $Fm3m$ [3, 4]. Ferrites are ferrimagnetic compounds of iron oxides Fe_2O_3 and FeO with general formula: $Fe^{3+}\left[Fe^{2+}_{1-y}Fe^{3+}_{1-y}Fe^{3+}_{1.67y*0.33y}\right]O_4$ (*indicates vacancies), which can be partly changed by other transition metal (TM) oxides [5]. The ferrites can be classified according to their crystalline structures: hexagonal ($MFe_{12}O_{19}$), garnet ($M_3Fe_5O_{12}$) and spinel (MFe_2O_4), where M = TM = Mn, Fe, Co, Ni, Cu, Zn. Both Heusler alloys and nanoferrites are important magnetic materials owing to their promise to qualify for many potential applications.

1.1 Heusler alloys

Heusler alloys have attracted avid attention due to their various properties suitable for technological applications such as high-density magnetic recording or magneto-optics applications [6–8]. It has been shown that the ordered combination of two binary B2 compounds XY and YZ leads to the formation of Heusler structure. Both these compounds may have CsCl type crystal structure e.g. CoAl and CoMn yield Co_2MnAl [9]. Thus the possible formation of new Heusler alloys will depend upon the ability of compounds to form B2 structure. It has also been observed to leave one of the four sublattices unoccupied ($C1_b$ structure). The $L2_1$ compounds are known as full-Heusler alloys while the latter compounds are often called half- or semi-Heusler alloys (**Figure 1**). Most of the Heusler alloys are saturate in weak magnetic field and ordered ferromagnetically. There are various parameters which are found to be very important in determining the magnetic properties [6–9, 11–13]; these include crystal structure, composition and heat treatment. The magnetic shape memory effect as well as other properties arises due to magneto-crystalline coupling in Heusler alloys [6–9, 11]. These made Heusler alloys to possess very interesting magnetic properties. A series of interesting diverse magnetic phenomenon like itinerant and localized magnetism, antiferromagnetism, helimagnetism, Pauli paramagnetism or heavy-fermionic behaviour can be studied in the same family of Heusler alloys [3, 4, 6, 7]. At low temperatures several Heusler alloys, e.g. Ni_2MnGa, Co_2NbSn etc., undergo a martensitic transition from a highly symmetric cubic austenitic to a low symmetry martensitic phase. Unlike atomic order–disorder transitions a martensitic transition is caused by non-diffusional

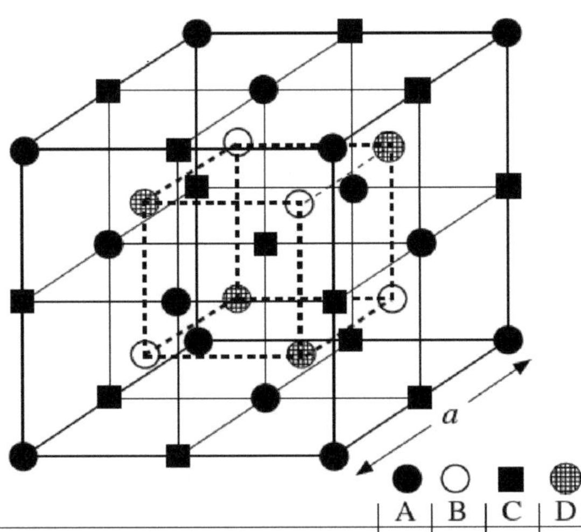

	A	B	C	D
Semi–Heuslers ($C1_b$) : XYZ	X	Y		Z
Full–Heuslers ($L2_1$) : X_2YZ	X	Y	X	Z
Quaternary Heuslers : (XX')YZ	X	Y	X'	Z
Inverse Heuslers (XA) : X_2YZ	X	X	Y	Z

Figure 1.
Representation of various structures of semi- and full-Heusler alloys. The lattice is consisted to have four interpenetrating f.c.c. lattices in all cases. It should be noted that if all atoms are identical, the lattice would simply become b.c.c [10].

cooperative movement of the atoms in the crystal. Heusler alloys exhibit magnetic shape memory effect (MSM), when they are magnetic in the martensitic phase [14–16]. This occurs especially in those cases when the Y constituent is Mn, but other transition elements are also possible. When these alloys are applied in the martensitic state, an external magnetic field can induce large strains.

In recent years, Ni-Mn-X (X = In, Sn, Sb) Heusler alloys have attracted much interest because of their magnetic shape memory effect and magnetic field-induced martensitic transformation [17–19]. In recent years, the interest in Heulser compounds has been increased due to the evolving field of spin-electronics [7]. These compounds have also been studied as potential spin-injector materials [6, 8, 20]. In the alloys X_2YZ, a value close to $4\mu_B$ is usually observed if the magnetic moment is carried by Mn atoms. These compounds are considered as ideal model systems to study the effects of atomic disorder and changes in the electron concentration on the magnetic properties. These compounds have localized magnetic properties though they are metals. Extensive magnetic and other measurements have been performed on quaternary Heusler alloys in order to understand the role of 3d (X) and sp. (Z) atoms on magnetic properties [21]. It has been shown that both the magnetic moment formation and the type of the magnetic order get influenced by the sp. electron concentration which is very important in establishing the magnetic properties. During the past few years, the family of ferromagnetic Heusler alloy systems has been extensively studied due to their main advantages in comparison to half-metallic systems e.g. structural similarity with the binary semiconductors and predicted perfect spin polarization at the Fermi level as well as high Curie temperature [22, 23]. The effect of alloying addition on the magnetic properties has been studied in Heusler alloys [24–30].

1.2 Nanoferrites

A series of spinel-structured ferrites, MFe_2O_4, where M = TM = Mn, Fe, Co, Ni, Cu, Zn were reported for novel data storage, recording devices, microwave technology and biomedical applications [31–35]. The spinel structure has general formula AB_2O_4 which had octahedral (B) and tetrahedral (A) sites. The spinel is normal, if M^{2+} occupies only the A sites; the spinel is inverse, if it occupies only the B sites. When Mn^{2+}occupies both A and B sites, the MFe_2O_4 has formed a mixed spinel structure while the other metal ferrites have an inverse spinel structure [3]. The magnetic spins at B sites align in parallel in direction to the applied magnetic field, whereas those present at A sites align antiparallel. Spinel ferrite has general formula: $\left(A^{2+}\right)\left[B_2^{3+}\right]O_4^{2-}$; where A^{2+} and B_2^{3+} are the divalent and trivalent cations occupying tetrahedral (A) and octahedral [B] sites. **Figure 2** shows the inverse spinel structure of $NiFe_2O_4$ [36]. The inverse spinel has general formula: $\left(M_{1-x}^{2+}Fe_x^{3+}\right)\left[M_x^{2+}Fe_{2-x}^{3+}\right]\left\{O^{2-}\right\}_4$; the round and square brackets denote (A) and [B] sites respectively, whereas x represents inversion degree. Also for normal spinel AB_2O_4, the A^{2+}cations occupies 1/8 of the *fcc* tetrahedral sites (T_d) while the B^{3+}occupy 16 of the 32 available octahedral sites (O_h). An inverse spinel structure with Ni^{2+} at octahedral sites (labelled as $Ni(O_h)$) and Fe^{3+} equally distributed between octahedral ($Fe(O_h)$)and tetrahedral sites ($Fe(T_d)$) of the O^{2-} fcc cell have been adopted for $NiFe_2O_4$ (**Figure 2**). The complete structure crystallizes into a cubic system O_h^7 with space group 227. The oxygen atoms occupy the 32e positions, $Fe(T_d)$ atoms occupy the 8a ones while the $Ni(O_h)$ and $Fe(O_h)$ atoms are distributed on the 16d positions, using Wyckoffnotations [36]. The efficacy of the material depends on its microstructural properties that are sensitive to mode of preparation. Moreover, $CoFe_2O_4$ is the most versatile hard ferrite with mixed cubic spinel

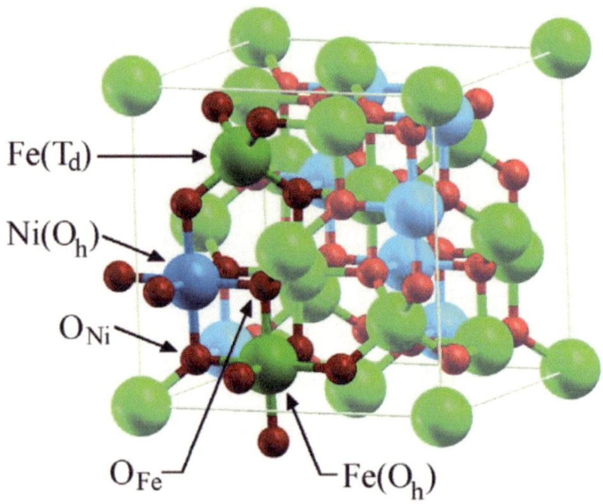

Figure 2.
Inverse spinel unit cell of $NiFe_2O_4$ Oxygen atoms are adjustable to complete the $Ni(O_h)$ and $Fe(O_h)$. Ni atoms (blue) Fe (green) and O atoms (red) [36].

structure having Fd3m space group and exhibits high coercivity ~5400 Oe, high magneto-crystalline anisotropy and moderate saturation magnetization [37, 38].

Spintronics is concerned with highly spin-polarized materials to enhance tunnelling magnetoresistance of magnetic tunnel junctions which are active members of magnetic random access memory elements. For the optimal operation of spintronics, the highly spin-polarized materials used for increasing spin-polarization of currents injected into semiconductors [38]. There is a way to achieve high spin-polarization by employing fully spin-polarized ferromagnetic metals, such as half-metals. Another way is to exploit the band structure features of tunnel barrier materials, *i.e.*, MgO, and filtering electronic wave functions. Spinel $NiFe_2O_4$, $CoFe_2O_4$, and $MnFe_2O_4$ are also used as such spin filters. Spin-dependent gap should result in spin-dependent barrier for tunnelling of electrons through the insulator, giving rise to spin filtering. In particular, a spin filtering efficiency of up to 22% is reported for $NiFe_2O_4$ barrier [39]. The resistive switching performance characteristics of a $Pt/NiFe_2O_4/Pt$ structure such as low operating voltage, high device yield, long retention time (up to 10^5s), and good endurance (up to 2.2×10^4 cycles) can be used to demonstrate the opportunity of spinel ferrites in non-volatile memory devices [33]. Since the resistive switching memory cell has capacitor-like metal/insulator/metal configuration, which can be switched reversibly between two different resistance states, i.e., high-resistance state (HRS) and low-resistance state (LRS).

Figure 3(a) shows the current–voltage (I-V) characteristics of the $Pt/NiFe_2O_4/Pt$ devices during repetitive switching cycles. Unipolar resistive switching characteristics were clearly observed in both the forward- and backward-bias sweeping processes. The drastic decrease in the current has been observed on increasing the forward voltage to its critical value of 0.6–1.0 V. This indicates the switching of the $Pt/NiFe_2O_4/Pt$ device from the LRS to the HRS. The device was switched from HRS to LRS, when soft breakdown occurred on increasing the voltage in the range of 1.8–2.2 V. $NiFe_2O_4$ has the inverse spinel structure which has Fe-O bonds that are stronger than Ni-O bonds leads to oxygen vacancies formation [40]. **Figure 3(b)** shows magnetic reduction effect in changing the valence of Fe^{3+} and Ni^{2+}ions [41]. The oxygen vacancies and the reduction of cations may cause the decrease in magnetization and the increase

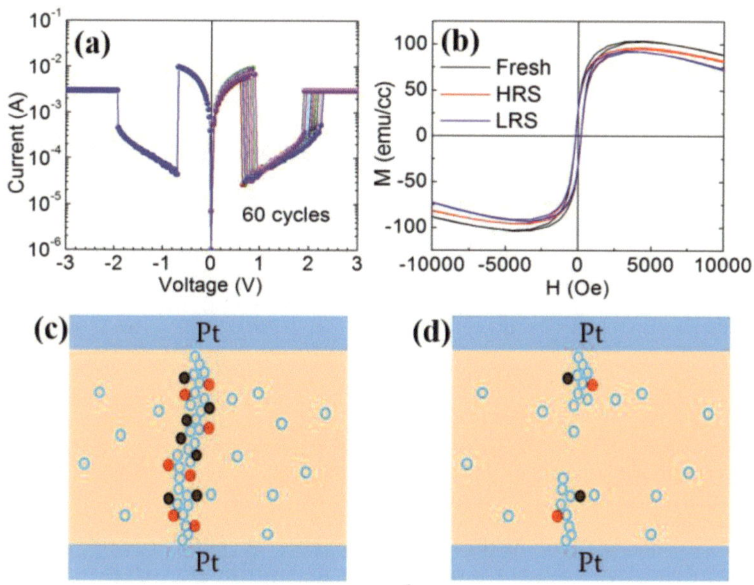

Figure 3.
(a) I-V characteristics with switching cycles of Pt/$NiFe_2O_4$/Pt devices (b) Magnetic hysteresis at 300 K for HRS and LRS (c & d) Conducting filaments mechanism in LRS and HRS [33].

in the electrical conductivity. Due to the annihilation of oxygen vacancies (driven by the thermal effect) and the change in the valence of cations (due to redox effect in the reset process) has led to rupture of filaments. The process of formation and rupture of the conducting filaments in $NiFe_2O_4$ films has been described in **Figure 3(c)** and **(d)**.

2. Structure/microstructure and magnetic properties of Heusler alloys

The $Cu_{50}Mn_{25}Al_{25-x}Ga_x$ (x = 0, 2, 4, 8 and 10) alloys with a thickness of ~40 to 50 μm and lengths of ~1 to 2 cm has been synthesized using melt spinning technique [24]. The substitution of Ga in place of Al is of special interest as it does not significantly change the lattice constant and the valence electron ratio (e/a) of the investigated alloy system [42–49]. The effect of heat treatment on the magnetic and phase transformation behaviour of $Cu_{50}Mn_{25}Al_{25-x}Ga_x$ ribbons will be discussed. The study is focussed on the structural/microstructural changes with substitution of Ga and their correlation with magnetic properties of $Cu_{50}Mn_{25}Al_{25-x}Ga_x$ alloys.

2.1 Microstructural and structural features

The effect of Ga substitution on the stability of Heusler phase has been investigated. The XRD patterns of melt-spun $Cu_{50}Mn_{25}Al_{25-x}Ga_x$ (x = 0, 2, 4, 8 and 10) alloys are shown in **Figure 4(a)**. A single Heusler phase of the Cu_2MnAl structure (space group: *Fm*3*m*, *a* = 5.949 Å) was observed for the alloys with x = 0, 2, 4 and 8, while the alloy with x = 10 reveals some diffraction peaks corresponding to γ-Cu_9Al_4 type phase (Space group: *P*43*m*, *a* = 8.702 Å). Thus it can be said that the Heusler phase was also observed in the Ga substituted alloys and is stable up to x = 8. In addition to the formation of Heusler phase, the precipitation of additional crystalline phase of γ-Cu_9Al_4 type has been observed for the alloys with x > 8. For the alloys from x = 0 to x = 8, a slight increase in the lattice parameter was observed. This is

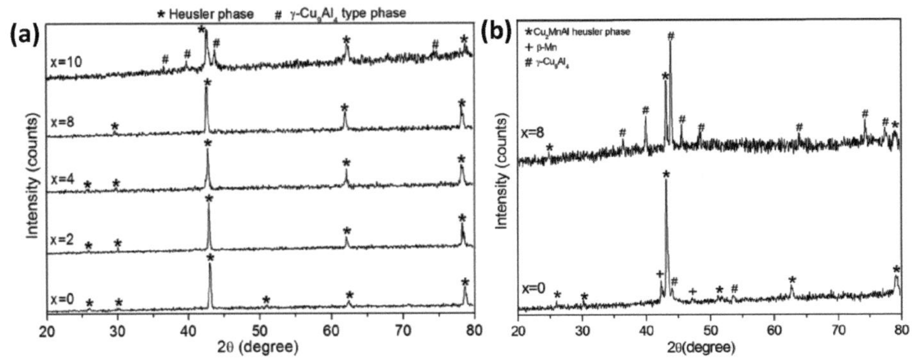

Figure 4.
(a) XRD patterns of $Cu_{50}Mn_{25}Al_{25-x}Ga_x$ (x = 0, 2, 4, 8 and 10) melt-spun ribbons (b) XRD patterns of $Cu_{50}Mn_{25}Al_{25-x}Ga_x$ ribbons annealed at 903 K for 30 h (Reprinted from reference [24] with kind permission from Elsevier, Copyright 2012, Elsevier).

expected due to nearly equal atomic size of Ga and Al. **Table 1** shows the variation of lattice parameter with Ga content. There is no significant increase in the lattice parameter for x = 10 (**Table 1**). The formation of extra phases for concentrations x > 8 may be due to the excess of Ga atoms which no longer replace the corresponding number of Al atoms in the crystal structure. Thus, the lattice constant of the Heusler phase remains almost constant. It is therefore can be concluded that the Ga atoms can substitute for Al atoms in the structure only within the content range of $0 \leq x \leq 8$. Hence, x = 8 is the critical Ga concentration (x_c), beyond which the alloys are having mixed phases.

The phase transformation behaviour of $Cu_{50}Mn_{25}Al_{25-x}Ga_x$ alloys on heat treatment has been investigated. The alloys with x = 0 and x = 8 are annealed at 903 K for 30 hrs and there XRD patterns are shown in **Figure 4(b)**. The XRD pattern for x = 0 shows the presence of Cu_2MnAl, β-Mn and γ-Cu_9Al_4 phases. The decomposition reaction of the Cu_2MnAl phase during the annealing process leads to the appearance of these two phases. Based on the previous studies done on the decomposition process of Cu-Mn-Al alloys, it has been shown that the Cu_2MnAl phase is metastable and can decompose into β-Mn and γ-Cu_9Al_4 phases at annealing temperatures from 800 to 900 K [50, 51]. This decomposition then follows the reaction ($Cu_2MnAl \rightarrow$ β-Mn + γ-Cu_9Al_4). However, the x = 8 alloy shows different decomposition behaviour. The Cu_2MnAl phase decomposes into only γ-Cu_9Al_4 on annealing. There are no traces of β-Mn phase here. In addition to this, Cu_2MnAl precipitated as a majority phase for the alloy with x = 0 while for the alloy with x = 8, γ-Cu_9Al_4 precipitated as a majority phase.

x (at %)	As-synthesized ribbons			Annealed ribbons	
	Lattice parameter a (Å)	M_s (emu/g) at 5 K	M_s (emu/g) at 300 K	M_s (emu/g) at 5 K	M_s (emu/g) at 300 K
0	5.947	95	85	53	45
4	5.960	88	79	—	—
8	5.986	83	73	5	3
10	5.981	20	16	—	—

Table 1.
The lattice constants, saturation magnetization (M_s) at 5 K and 300 K for melt-spun and annealed ribbons of $Cu_{50}Mn_{25}Al_{25-x}Ga_x$ (x = 0, 4, 8 and 10) alloys (Reprinted from reference [24] with kind permission from Elsevier, Copyright 2012, Elsevier).

Further TEM characterization studies of melt-spun as well as annealed samples were carried out. TEM microstructure and its corresponding diffraction pattern of $Cu_{50}Mn_{25}Al_{25}$ and $Cu_{50}Mn_{25}Al_{17}Ga_8$ melt-spun alloys are shown in **Figure 5** (on the left). The formation of nano-meter sized grains in the range of 100 to 200 nm of the alloys with x = 0 and x = 8 have been observed (*c.f.* **Figure 5(a)-(d)**). The substitution of Ga in $Cu_{50}Mn_{25}Al_{25-x}Ga_x$ alloy system does not results in any significant microstructural variation. The confirmation of Cu_2MnAl Heusler phase has been done based on the analysis of selected area diffraction pattern (SADP) of the alloys. The TEM microstructure and diffraction pattern for the annealed ribbon of x = 8 are shown in **Figure 5** (on the right). The SADP analysis reveals two crystalline phases; Cu_2MnAl and γ-Cu_9Al_4. This is consistent with the result obtained by XRD analysis which also shows the formation of Cu_2MnAl and γ-Cu_9Al_4 phases. The large grains belong to Cu_2MnAl phase while the small ones identified as γ-Cu_9Al_4 phase. As one might know, Cu_2MnAl phase is ferromagnetic while γ-Cu_9Al_4 phase is paramagnetic in nature [50].

2.2 Magnetic properties

The temperatures of 5 K and 300 K were used to measure the magnetization curves (M-H curves) of $Cu_{50}Mn_{25}Al_{25-x}Ga_x$ (x = 0, 4, 8 and 10) alloys. Based on the results presented in **Figure 6(a)** and **(b)**, It can be said that the magnetization is saturated in a magnetic field of about 1500 to 2500 Oe, indicating that ribbons are fairly homogenous ferromagnets [51]. The soft ferromagnetic behaviour (at 5 K and 300 K) were observed for the $Cu_{50}Mn_{25}Al_{25-x}Ga_x$ (x = 0, 4, 8 and 10) alloys. **Table 1** gives the list of the saturation magnetization (M_s) derived from the M-H curves for all the compositions. The M_s for the composition with x = 0 at 5 K is ~95 emu/g and at 300 K is ~85 emu/g respectively. These are very close to the reported magnetization values for this alloy [52]. It can be seen from **Table 1** that M_s decreases with increasing Ga content. This decrease is very small from x = 0 (~ 95 emu/g) to x = 8 (~ 83 emu/g) at 5 K. However, for x > 8, the M_s decreases from ~83 emu/g (for x = 8) to ~20 emu/g (for x = 10). The large decrease in the M_s for x = 10 may be explained on the basis that the Cu_2MnAl Heusler phase which is responsible for the ferromagnetism is stable only up to x = 8. The decomposition of Heusler phase into γ-Cu_9Al_4 type crystalline phase for the alloys with x > 8, is responsible for the significant decrease in the M_s for x = 10. The M_s values in the range 95–83 emu/g (at 5 K) has

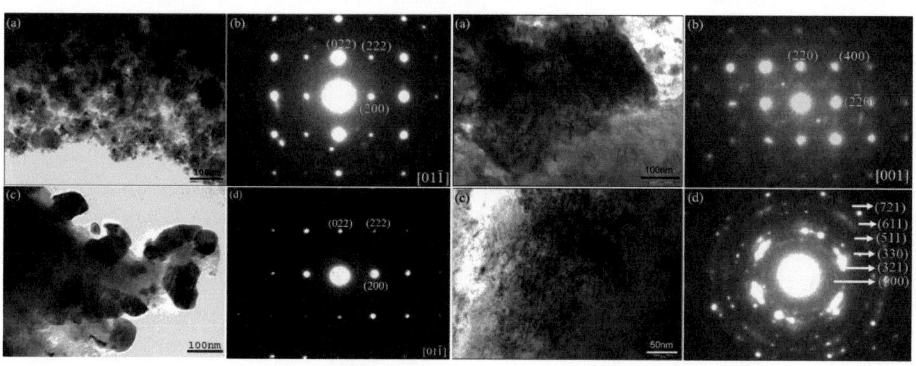

Figure 5.
(left) TEM microstructures and the corresponding selected area diffraction patterns of the melt-spun alloys with x = 0 (a and b) and x = 8 (c and d) showing the formation of Cu_2MnAl Heusler phase. (Right) TEM microstructures and the corresponding selected area diffraction patterns of the annealed alloy of x = 8 showing the existence of (a and b) Cu_2MnAl and (c and d) γ-Cu_9Al_4 phases (Reprinted from reference [24] with kind permission from Elsevier, Copyright 2012, Elsevier).

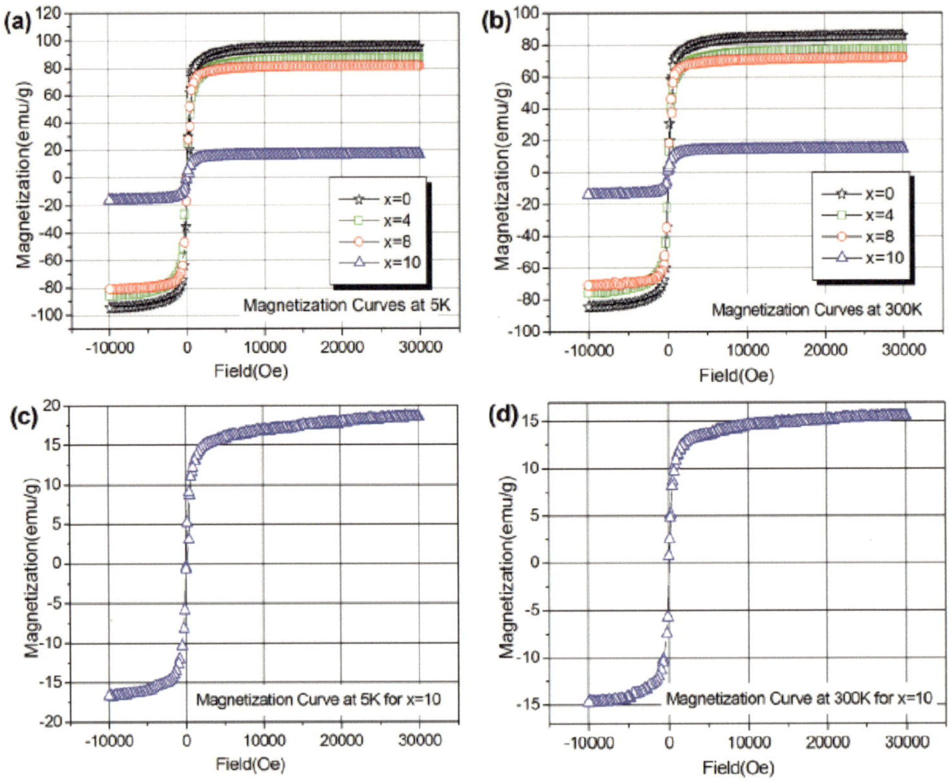

Figure 6.
Magnetization curves at 5 K (a) and 300 K (b) for melt-spun ribbons of $Cu_{50}Mn_{25}Al_{25-x}Ga_x$ (x = 0, 4, 8 and 10) alloys. Magnetization curves (c) and (d) shows a small increase in M with H for the alloy with x = 10 (Reprinted from reference [24] with kind permission from Elsevier, Copyright 2012, Elsevier).

been observed for the alloys with x = 0, 4 and 8 while the M_s value of 20 emu/g (at 5 K) has been observed for x = 10. Thus there is very slight variation in the M_s for x = 0, 4 and 8. The slight decrease in the M_s for x = 4 and x = 8 as compared to x = 0 may be due to several reasons. In the present alloy system, Ga is a non-magnetic element. The concentration of Al is only varied while the concentration of Cu and Mn remains fixed in $Cu_{50}Mn_{25}Al_{25-x}Ga_x$ alloy system. It has been reported that Mn-Mn coupling is responsible for the magnetic properties of $Cu_{50}Mn_{25}Al_{25}$ alloy [52, 53]. As discussed earlier, the substitution of Ga may increase the Mn-Mn distance in the lattice. This is due to sight increase in the lattice constant (see **Table 1**). Thus, the decrease in the magnetization has been observed due to increase in the lattice constant which reduces the ferromagnetic coupling of $Cu_{50}Mn_{25}Al_{25-x}Ga_x$ alloy system. Such type of observation has also been reported in other alloy systems [14, 26, 29, 30]. Also, it is well known that Ga^{3+} is normally substituted in isovalent state vis-a-vis that of Al^{3+} [54–59]. However, it is worth mentioning here that Ga^{1+} may possess monovalent state. It has been confirmed Al, Ga mainly provide conduction electrons for the exchange interaction and the Mn atoms show a localized moment in this material [3]. It may be said by considering the varied concentration ratio of Al and Ga that the concentration of conduction electron plays a very important role in determining the magnetic properties. Thus, the magnetic properties may be affected by the presence of mixed valence states of Ga. The concentration of Ga and the disordered occupation of Mn atoms may also be important factors which affects the magnetic properties. Therefore, the conduction electron concentration is believed to be critical in stabilizing the Heusler structure [3]. In the present case, the concentration of Mn is fixed i.e. 25 at. %. Thus, the ferromagnetism behaviour of this material may be related to the concentration of Mn, atomic sites and situation of conduction electrons provided by Cu, Al and Ga. Further, the M-H curves for the melt spun alloy with x = 10 at 5 K and 300 K respectively are shown in **Figure 6(c)** and **(d)**. A close observation of the curves indicates a small increase in M with H superimposed on the flat saturated ferromagnetic M-H curve. This may be due to the formation of a paramagnetic phase (γ-Cu_9Al_4) in addition to the majority Cu_2MnAl ferromagnetic phase in the alloy with x = 10.

The effect of heat treatment on the magnetization has also been studied. On annealing, the magnetic properties of the alloy have changed. The hysteresis loops of the annealed ribbons (for x = 0 and x = 8) at 5 K and 300 K are shown in **Figure** 7. The M_s of the annealed ribbons are found to be lower than their respective as-synthesized ribbons (**Table 1**). The M_s ~ 53 emu/g has been found for annealed ribbon of x = 0 from the magnetization curve (M-H) at 5 K, which is lower than that of the value found for the as-synthesized ribbons of x = 0 *i.e.* M_s ~ 95 emu/g. The decomposition of Cu_2MnAl ferromagnetic phase into β-Mn and γ-Cu_9Al_4 paramagnetic phases may decrease the magnetization [50]. However, along with these phases, Cu_2MnAl Heusler phase is still a majority phase for the alloy with x = 0. The drastic decrease in the magnetization value at 5 K from ~83 emu/g to ~5 emu/g has been observed for x = 8. This is attributed to the different decomposition behaviour of x = 0 and x = 8 alloys. The M-H curves of the annealed ribbon for x = 8 (at 5 K and 300 K) are shown in **Figure** 7**(c)** and **(d)** (c & d). The increase in M with increase in the applied field is

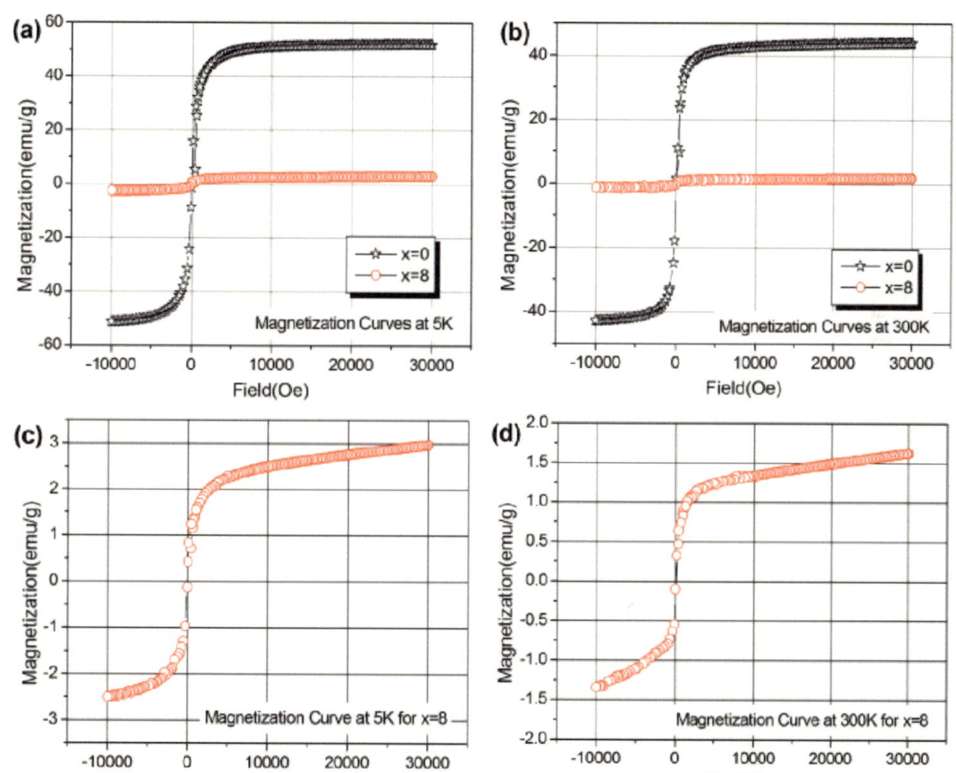

Figure 7.
Magnetization curves at 5 K (a) and 300 K (b) for annealed ribbons of $Cu_{50}Mn_{25}Al_{25-x}Ga_x$ *(x = 0 and 8) alloys. Magnetization curves (c) and (d) shows an increase in M with H for the alloy with x = 8 (Reprinted from reference [24] with kind permission from Elsevier, Copyright 2012, Elsevier).*

clearly evident. This may be due to the precipitation of paramagnetic phase (γ-Cu_9Al_4 type) as a majority phase for the alloy with x = 8.

The temperature dependence of magnetization curves (M-T curves) for the melt-spun and annealed ribbons under a field of 500 Oe has been investigated. **Figure 8(a-d)** shows the M-T curves of the melt-spun $Cu_{50}Mn_{25}Al_{25-x}Ga_x$ (x = 0, 4, 8 & 10) alloys. The alloys with x = 4 and 8 have the same type of M-T curves as the parent alloy $Cu_{50}Mn_{25}Al_{25}$ (x = 0) shows the characteristics of a ferromagnet. With increase in Ga concentration, the saturation magnetization M(T) decreases. At low temperature in the M-T curves for x = 0, 4 and 8, it follows the expression $M(T) = M(0)(1\text{-}AT^n)$ with n = 3/2 [52]. The data for T < 100 K as a function of $T^{3/2}$ is shown in **Figure 8(a)–(c)**. The $T^{3/2}$ dependence of the magnetization at this low T range has been shown by the solid line in the inset graph which is a linear fit to the data. The magnetization measurements of $Cu_{50}Mn_{25}Al_{25-x}Ga_x$ (x = 0, 4, 8 & 10) alloys in the zero field cooled (ZFC) and field cooled (FC) modes are shown in **Figure 8**. The magnetization measurements in ZFC and FC modes can give an estimate of magnetic ordering temperature in a system with long range magnetic order or blocking/freezing temperature in the case of medium range order [60].

The splitting in the ZFC and FC magnetization curves has been observed for the alloys with x = 0, 4, 8 and 10 (**Figure 8**). The alloys with x = 0, 4, 8 and 10 shows the splitting at ~145 K, ~125 K, ~25 K and ~ 225 K respectively. The splitting temperature found to be decreases from x = 0 to x = 8 and then increases for x = 10. The reason for this may be attributed to the fact that the alloys up to x = 8 exhibit single phase while the alloy with x = 10 is biphasic in nature. Thus, the synthesis acquired anisotropy

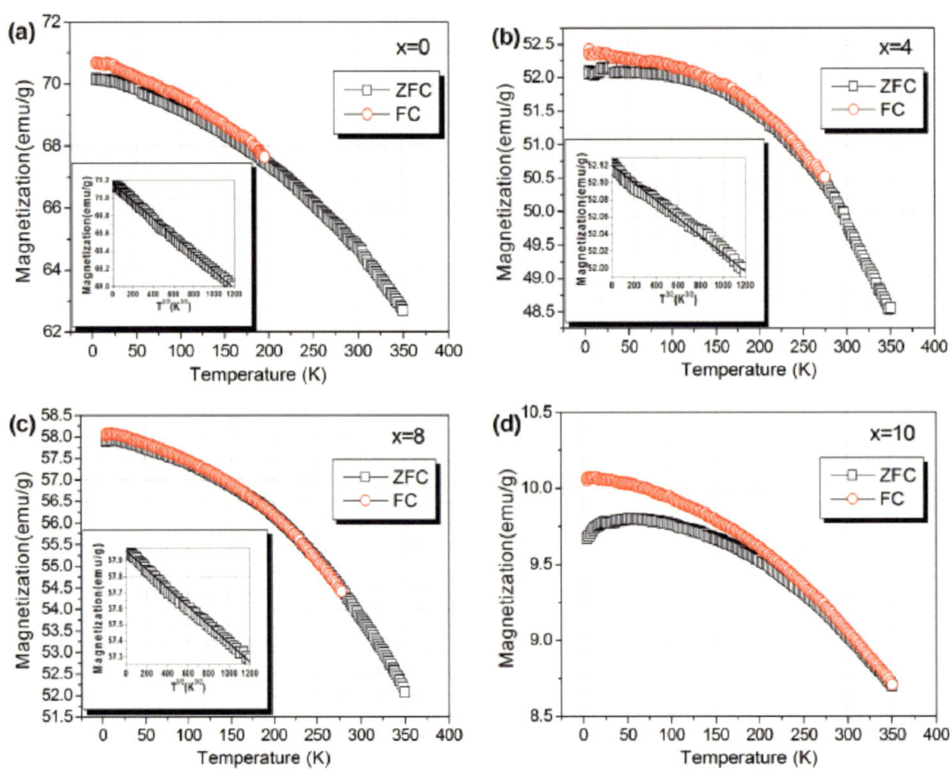

Figure 8.
Temperature dependence of magnetization for melt-spun ribbons of $Cu_{50}Mn_{25}Al_{25-x}Ga_x$ alloys with x = 0 (a), x = 4 (b), x = 8 (c) and x = 10 (d) in a field of 500 Oe in the ZFC and FC modes. The data for T < 100 K as a function of $T^{3/2}$ is shown in the inset graphs of (a), (b) and (c) (Reprinted from reference [24] with kind permission from Elsevier, Copyright 2012, Elsevier).

of the $Cu_{50}Mn_{25}Al_{25-x}Ga_x$ (x = 0–10) melt spun alloys causes the splitting in the ZFC and FC curves. The synthesis of ribbons using the melt spinning process involves the sudden cooling of molten alloy when it falls on to the rotating Cu-wheel. The sudden and fast cooling of the molten alloy arises from the very high speed (~ 4000 rpm) of the rotating wheel. It is well known that the side of the ribbon which is in direct contact with the wheel has higher cooling rate than the one which is not in contact [61]. Thus, the variation in cooling rate along the thickness of the ribbon result in some microstructural changes between contact side (CS) and non-contact side (NCS) [61]. This has led to synthesis acquired anisotropy in the ribbons.

3. Structure/microstructure and magnetic properties of nanoferrites

3.1 XRD and AFM analysis of $NiFe_2O_4$, $CoFe_2O_4$ and $MnFe_2O_4$ thin films

The nanoferrites of $NiFe_2O_4$, $CoFe_2O_4$ and $MnFe_2O_4$ thin films were prepared by a MOD method [62]. **Figure 9(a)** shows the XRD pattern of $NiFe_2O_4$, $CoFe_2O_4$ and $MnFe_2O_4$ thin films. The ferrite peaks at 2θ = 30.481, 34.991, 37.481, 42.581, 48.021, 51.231 and 55.85° respectively attributed to (220), (311), (222), (400), (331), (422) and (511) reflections of spinel structure. A small amount of α-Fe_2O_3 phase had also been formed as an impurity phase. The lattice constants are calculated from the XRD data *i.e.* a(Å) ~8.161, 8.312 and 8.425, respectively for $NiFe_2O_4$, $CoFe_2O_4$ and $MnFe_2O_4$. These calculated values of lattice constants are closer to bulk $NiFe_2O_4$

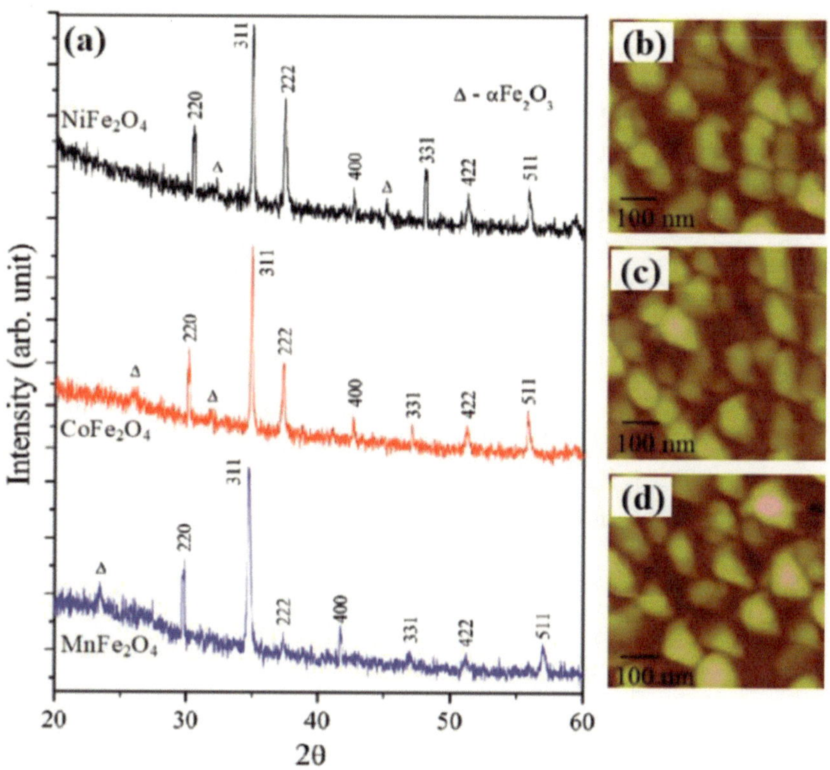

Figure 9.
(a) XRD pattern of $NiFe_2O_4$, $CoFe_2O_4$ and $MnFe_2O_4$ thin films. AFM images (b) $NiFe_2O_4$ (c) $CoFe_2O_4$ (d) $MnFe_2O_4$ thin films (Reprinted from reference [62] with kind permission from Elsevier, Copyright 2011, Elsevier).

(8.17 Å), $CoFe_2O_4$ (8.32 Å) and $MnFe_2O_4$ (8.43 Å) [63, 64]. **Figure 9(b)-(d)** shows the AFM images respectively of $NiFe_2O_4$, $CoFe_2O_4$ and $MnFe_2O_4$ thin films. It exhibited homogeneous micro-structures with uniform size distribution of nano-grains. The thin film surfaces are smooth and crack-free. The grains exhibit a round shape with a small grain boundary region. The average size of grains AFM is 46, 61 and 75 nm and surface roughness is 2.5, 4 and 2 nm respectively for $NiFe_2O_4$, $CoFe_2O_4$ and $MnFe_2O_4$. The smaller nano-grains formation may attribute better stoichiometric ratio and low processing temperature of crystallization. Also the addition of PEG encapsulates the ferrites constituents into smaller groups during the heating process [65].

3.2 Magnetism of ferrite MFe_2O_4 nanoparticles

3.2.1 Ferromagnetism of MFe_2O_4 [M = Mn, Co, Ni, Zn] with $BaTiO_3$ nanocomposite

The magnetic nanoparticles of multiferroic $MFe_2O_4/BaTiO_3$ [M = Mn (MnFO/BTO), Co (CFO/BTO), Ni (NFO/BTO) and Zn (ZFO/BTO)] thin films were fabricated by a MOD method [66]. The addition of ferrite MFe_2O_4 in perovskite $BaTiO_3$ results in to lattice strain due to tetragonal distortion, expansion/contraction of MFO/BTO unit cell and lattice mismatch. The tetragonal BTO, spinel MFO phases and lattice strain effects are confirmed by XRD analysis. The average grain size for MnFO/BTO, CFO/BTO, NFO/BTO and ZFO/BTO is calculated from AFM images and is found to be 25, 102, 24 and 133 nm, respectively.

The ferromagnetic behaviour (magnetization versus applied magnetizing field (M-H_{dc})) of MFO/ BTO thin films at room temperature is shown in **Figure 10(a)**. The values of remanent magnetization, M_r = 0.03, 3.75, 6.76 and 1.49 k J T^{-1} m^{-3} with coercivity, H_c = 0.013 × 10^5, 0.079 × 10^5, 0.167 × 10^5 and 0.135 × $10^5$$Am^{-1}$ and saturation magnetization, M_s = 1.29, 20.25, 27.64 and 6.77 kJ T^{-1} m^{-3}, respectively, measured for MnFO/ BTO, CFO/ BTO, NFO/ BTO and ZFO/ BTO. The magnetization values observed for MFO/BTO nanocomposite shown abrupt reduction than single phase $MnFe_2O_4$ (5.40 kJ T^{-1} m^{-3}), $ZnFe_2O_4$ (230 kJ T^{-1} m^{-3}), $NiFe_2O_4$ (50.60 kJ T^{-1} m^{-3}) and $CoFe_2O_4$ (33.50 kJ T^{-1} m^{-3}) [62, 67]. Due to non-magnetic

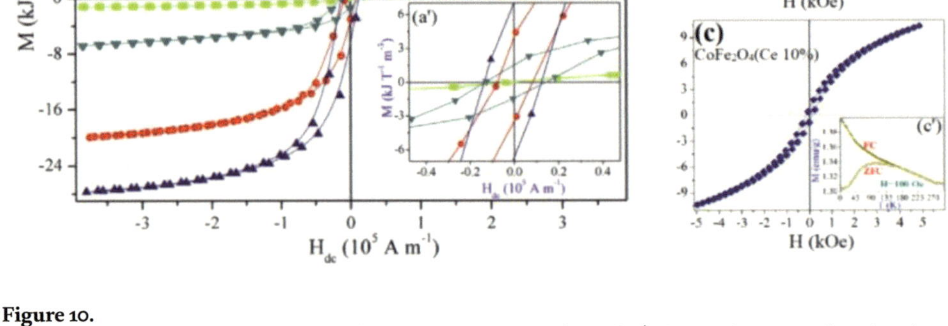

Figure 10.
(a) Room temperature M-H_{dc} hysteresis of $MFe_2O_4/BaTiO_3$ thin films (b & c) M-H hysteresis of Ce doped $CoFe_2O_4$ nanoparticles at room temperature. Inset shows SQUID measurement for ZFC/FC experiment (Reprinted from reference [66] with kind permission from Elsevier, Copyright 2017, Elsevier).

BTO phase of nano-composite, the saturation magnetization of ferrite reduces. Since mixed perovskite BTO acts as a non-magnetic defect, which hinders the growth of the magnetic domains of spinel MFO and their movement under an external magnetic field [68]. Thus the non-magnetic elements weaken the A-B super exchange interaction which results in the increase of distance between the magnetic moments in A and B sites in the spinel structure [69]. The weaker A-B super exchange interaction of MFO by non-magnetic BTO phase is more affected by the thermal motion, resulting in the decrease of T_C. Wang *et al.* [70] suggested that the decrease in the magnetization causes the relaxation process to increase, which might be related with oxygen vacancies redistribution. Bateer *et al.* [71] suggested that the decrease in particle size and the presence of a magnetic dead oranti-ferro-magnetic layer on the surface results into reduction in saturation magnetization in ferrite nanoparticles.

Moreover the origin of observed ferromagnetic behavior (**Figure 10**) of MFO ferrites nanocomposite must be related with two different mechanisms: ferrimagnetic coupling of Fe ionsat A-B sites in $(M_{1-x}Fe_x)[M_xFe_{2-x}]O_4$ and the surface spin canting [72]. The structure of the ferrite becomes mixed spinel from its normal configuration when consider the cation inversion in which the shifting of M ions from A to B site and that of Fe ions from B to A-site changes the magnetic behavior of the nanocomposite [73]. The magnetic moments of Fe ions (at B-site) cancel out due to negative interaction among them and M ions have nothing to contribute. The Fe ions present at A-site are responsible for magnetization enhancement and contribute to net magnetic moment. Vamvakidis *et al.* [74] suggested that within the spinel structure, the reduction in inversion degree (~0.22) of $MnFe_2O_4$ (due to the partial oxidation of Mn^{2+} to Mn^{3+} ions) results into weaker super exchange interactions between tetrahedral and octahedral sites. It contributes to weak ferrimagnetism in $MnFe_2O_4$. Bullita *et al.* [75] reported that the cation distribution of $ZnFe_2O_4$ at the nanoscale level is contributed by partial inverted spinel structure which results in the increase of magnetization. Peddis *et al.* [76] shows the typical ferrimagnetic structure of inverse $CoFe_2O_4$ nanoparticles with an inversion degree of 0.74. These results in the better correlation between spin canting and cationic distribution to get competitively higher value of saturation magnetization. However, Carta *et al.* [77] reported the degree of inversion 0.20 for $MnFe_2O_4$ nanoparticles, 0.68 for $CoFe_2O_4$ nanoparticles and 1.00 for $NiFe_2O_4$ nanoparticles.

3.2.2 Ferromagnetic ordering in Ce doped $CoFe_2O_4$ nanoparticles

It is reported that the core/shell nanoparticles of $CoFe_{2-x}Ce_xO4$ [x = 0.05 (CFCeO05), 0.1 (CFCeO10)] were prepared by a chemical combustion method [37]. XRD pattern results into spinel structure with cubic space group. From TEM images, the average particles size, D = 8 and 10 nm, respectively, for CFCeO05 and CFCeO10 sample. **Figure 10(b)** and **(c)** shows the M-H hysteresis with M_s = 42.54 and 10.41 emug^{-1} and M_r = 26.68 and 1.57, emug^{-1} with H_c = 1526 and 140 Oe, respectively, observed for CFCeO05 and CFCeO10. These values of magnetization are larger than nanostructured pure $CoFe_2O_4$ but smaller than bulk value (73 emug^{-1} at room temperature). This is due to higher surface energy and surface tension in the nanoparticles which changes cationic preferences of ferrite. It leads to an increase in degree of anti-site defects to cause more surface spin canted or disorder.

Recently, Georgiadou *et al.* [78] suggested modification in cation occupancy in nanostructured CFO due to its inverse spinel structure. The theoretical expression for the net moment of M_{Ce} is given (Ce^{3+} ions occupy only the B sites for their large ionic size): $M_{Ce} = M_B - M_A = \left[5\times(1-x)+3.8+x\times\mu_{Ce}\right]-5$; where M_A and M_B are the wof Fe^{3+}cation is fixed to $5\mu_B$ (spin only) and for octahedrally coordinated Co^{2+} cations is fixed to 3.8 which correspond to the M_{sat} at 0 K of bulk CFO (95 $emug^{-1}$) [79]. The net magnetic moments μ_{Ce} cation is zero for the diamagnetic Ce^{4+} ions and non-zero for the paramagnetic Ce^{3+} ones. By replacing Fe^{3+} by Ce^{3+}cation, the M_s is expected to vary as μ_{Ce} by the sequential filling of electrons in the 4*f* shell. Unlikely, a clear deviation between the theoretically predicted magnetization by above equation and the experimental one (**Figure 10(b)** and **(c)**) is observed. This is ascribed into two reasons. One is the decrease in strong negative Fe^{3+}-Fe^{3+} interaction that resulted from the doping of Ce because the spinel ferrimagnetic CFO is largely governed by the negative Fe^{3+}-Fe^{3+} interaction (the spin coupling of the 3d electrons). The Ce^{3+}-Fe^{3+} interaction (4*f*-3*d* coupling) as well as the Ce^{3+}-Ce^{3+} one (indirect 4*f*-5*d*-4*f* coupling) exist this is very weak [80]. The other is the rearrangement of the Co^{2+} ion in the A and B sites resulted from the doping with RE Ce^{3+} ions. The migration of Co^{2+} ion into tetrahedral sublattice decreases the concentration of Fe^{3+} ion in A site, leading to enhance M_s. Also, the value of H_c is reduces with Ce^{3+} ion concentration in CFCeO10 and shows weak ferromagnetism. This is responsible due to variation in core/ shell formation and description of CFO lattice by vibrational modes [37]. The core/shell (CeO/CFO) system result into an increase in effective magnetic anisotropy caused by surface and interface exchange coupling effects. The huge difference in the coercivity value among CFCeO samples may ascribe to surface pinning that arises due to missing coordination of oxygen atoms and the shape effect of the spinel ferrite.

The origin of observed room temperature ferromagnetism of CFCeO samples is evaluated by the temperature dependent magnetization [M(T)] with field cooling (FC) and zero field cooling (ZFC) measurement (**Figure 10(b')** and **(c')** inset). The applied magnetic field is 100 Oe. These M-T measurements show that ZFC-FC curves of the CFCeO05 sample did not coincide with each other or slightly coincide around 300 K. It indicates that the nanoparticles are still magnetically blocked at around room temperature. However, CFCeO10 nanoparticles show blocking temperature of antiferromagnetism, spin glass etc. at about 91 K. This type of the magnetic response is due to different nano-core/ shell formation in CFCeO samples. The core/shell nanoparticles provides spin-phonon coupling in which a core of aligned spins is surrounded by a magnetically disordered shell [81].

3.2.3 Humidity response from $MgFe_2O_4$ thin films

Figure 11 shows the humidity response of $MgFe_2O_4$ thin films (measured in the range 10–90% RH at 25°C). The base resistance of thin film annealed at 400°C increased from 59 GΩ to 30 TΩ annealed at 800°C. There are many factors on which the resistance of ferrites depends such as porosity, vacancies and electron hopping between Fe^{2+} and Fe^{3+}. In the present study it may be due to higher annealing temperature that increases the average pore size distribution which further creates more obstruction for the charge carrier's movement. It can be seen from **Figure 11** that with higher annealing temperature, the response of Log R (Ohm) approaches towards linearity with rising humidity 10–90% RH (relative humidity). Log R variation of the film annealed at 400°C was almost constant (up to 50% RH), after that the linear decrease was observed at high humidity value. This may be due to the presence of less pores available for adsorption and thus a very few water molecules are only able to chemisorbed in such pores. The slope of log R (with increasing humidity) increases for the film annealed at 600°C and it became almost linear for the film annealed at 800°C for the entire range of humidity which may be due to the increase of intra and inter pores.

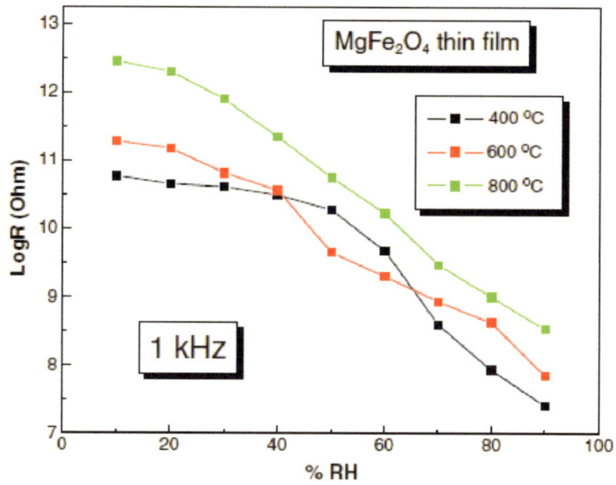

Figure 11.
Log R vs. relative humidity response curve of $MgFe_2O_4$ thin films at 1 kHz (Reprinted from reference [82] with kind permission from Elsevier, Copyright 2011, Elsevier).

4. Conclusion

In this chapter the recent progress in the development of Heusler alloys and nanoferrites are discussed. Formation of Heusler single phase of the Cu_2MnAl structure has been found only for the $Cu_{50}Mn_{25}Al_{25-x}Ga_x$ alloys with $x \leq 8$. The size of the grains for the Heusler phase alloy is in the range 100–200 nm. Long term annealing of the alloys leads to the formation of β-Mn and γ-Cu_9Al_4 type phases. The saturation magnetization (M_s) decreases with increasing Ga concentration. The decomposition of the Cu_2MnAl Heusler phase into β-Mn and γ-Cu_9Al_4 phases during annealing leads to the decrease in the magnitude of M_s.

The XRD pattern confirms the cubic phase of spinel ferrites in nanoferrites. The average grain's size for these ferrites is found to be less than 100 nm. Thus it can be said that the nanostructural formation of ferrite grains depends on the selection of chemical route and annealing temperature and the magnetic properties of nanoferrites are different from the bulk ferrites. The core/shell nanocomposites formation of Ce substituted $CoFe_2O_4$ hinders the superparamagnetism formed due to small magnetic nanoparticles results into long-range antiferromagnetic interactions. $MgFe_2O_4$ thin film annealed at 800°C has a linear log R (Ohm) response towards the entire humidity range 10–90% RH.

Acknowledgements

One of the authors (Devinder Singh) gratefully acknowledges the financial support by Department of Science and Technology (DST), New Delhi, India in the form of INSPIRE Faculty Award [IFA12-PH-39] and K.C. Verma is grateful to SERB Department of Science and Technology (DST), Govt. of India for the financial support under Fast-track Scheme for Young Scientist [SR/FTP/PS-180/2013].

Author details

Devinder Singh[1*] and Kuldeep Chand Verma[2]

1 Amity School of Applied Sciences, Amity University, Lucknow Campus, Lucknow, UP, India

2 CSIR-Central Scientific Instrument Organisation, Chandigarh, India

*Address all correspondence to: dsingh2@lko.amity.edu

References

[1] Heusler F, Starck W, Haupt E, Verh d. DPG 1903;**5**:220.

[2] Heusler F, Verh d. DPG 1903;**5**:219.

[3] Tobola J, Pierre J, Kaprzyk S, Skolozdra RV, Kouacou M.A. Electronic structure and magnetism in Co-Ni Heusler alloys. J Magn Mag Materials. 1996;**159**:192.

[4] Pierre J, Skolozdra RV, Tobola J, Kaprzyk S, Hordequin C, Kouacou MA, Karla I, Currat R, Leliévre-Berna E. Properties on request in semi-Heusler phases. J Alloys Compds. 1997;**262-263**:101.

[5] Santina Mohallem ND, Silva JB, Nascimento GLT, Guimaraes VL. Study of multifunctional nanocomposites formed by cobalt ferrite dispersed in a silica matrix prepared by sol-gel process. InTech, pp. 457 (Chapter 18).

[6] Cong DY, Roth S, Liu J, Luo Q, Potschke M, Hurrich C, Schultz L. Superparamagnetic and superspin glass behaviors in the martensitic state of $Ni_{43.5}Co_{6.5}Mn_{39}Sn_{11}$ magnetic shape memory alloy. Appl Phys Lett. 2010;**96**:112504.

[7] Knobel M, Nunes WC, Socolovsky LM, Biasi ED, Vargas JM, Denardin JC. Superparamagnetism and other magnetic features in granular materials: A review on ideal and real systems. J Nanosci Nanotechnol. 2008;**8**:2836.

[8] Marukame T, Ishikawa T, Matsuda K, Uemura T, Yamamoto M. High tunnel magnetoresistance in fully epitaxial magnetic tunnel junctions with a full-Heusler alloy $Co_2Cr_{0.6}Fe_{0.4}Al$ thin film. Appl Phys Lett. 2006;**88**:262503.

[9] Yilgin R, Oogane M, Yakata S, Ando Y, Miyazaki T. Intrinsic Gilbert Damping Constant in Co_2MnAl Heusler Alloy Films. IEEE Transactions on Magnetics 2005;**41**:2799.

[10] Galanakis I. Theory of Heusler and Full-Heusler alloys. C. Felser and A. Hirohata (eds.), *Heusler Alloys*. Springer Series in Materials Science 222, DOI 10.1007/978-3-319-21449-8_1.

[11] Ouardi S, Shekhar C, Fecher GH, Kozina X, Stryganyuk G, Felser C, Ueda S, Kobayashi K. Electronic structure of Pt based topological Heusler alloys with $C1_b$ structure and “zero band gap”. Appl Phys Lett 2011;**98**:211901.

[12] Singh D, Singh D, Srivastava ON, Tiwari RS. Microstructural effect on the low temperature transport properties of Ce-Al (Ga) metallic glasses. Scripta Materialia 2016;**118**:24.

[13] Singh D, Singh D, Tiwari RS. Effect of Ga substitution on low temperature transport and magnetic response of $Ce_{75}Al_{25}$ metallic glass. AIP Advances 2018;**8**:095222.

[14] Singh RK, Raja MM, Mathur RP, Shamsuddin M. Effect of Fe substitution on the phase stability and magnetic properties of Mn-rich Ni-Mn-Ga ferromagnetic shape memory alloys. J Magn Mag Materials 2011;**323**:574.

[15] Kaul SN, Annie D’ Santhoshini B, A bhyankar AC, Barquin LF, Henry P. Thermoelastic martensitic transformation in ferromagnetic Ni–Fe–Al alloys: Effect of site disorder. Appl Phys Lett 2006;**89**:093119.

[16] Li Y, Jiang CB, Liang T, Ma YQ, Xu HB. Martensitic transformation and magnetization of Ni–Fe–Ga ferromagnetic shape memory alloys. Scripta Materialia 2003;**48**:1255.

[17] Liu J, Woodcock TG, Schoerbaum N, Gutfleisch O. Influence of annealing

on magnetic field-induced structural transformation and magnetocaloric effect in Ni–Mn–In–Co ribbons. Acta Materialia 2009;**57**:4911.

[18] Karaca HE, Karaman I, Basaran B, RenY,ChumlyakovYI,MaierHJ.Magnetic Field-Induced Phase Transformation in NiMnCoIn Magnetic Shape-Memory Alloys-A New Actuation Mechanism with Large Work Output. Adv Funct Mater 2009;**19**:983.

[19] Kainuma R, Ito W, Umetsu RY, Oikawa K, Ishida K. Magnetic field-induced reverse transformation in B2-type NiCoMnAl shape memory alloys. Appl Phys Lett 2008;**93**:091906.

[20] C Biswas, Rawat R, Barman SR. Large negative magnetoresistance in a ferromagnetic shape memory alloy: $Ni_{2+x}Mn_{1-x}Ga$. Appl Phys Lett 2005;**86**:202508.

[21] Webster P J, Ziebeck K R A. Alloys and Compounds of d-Elements with Main Group Elements - Part 2. H. P. J. Wijn, Landolt-Bornstein, New Series, Group III, 19/c (Springer, Berlin), (1988).

[22] Groot RAD, Muller FM, Engen PGV, Buschow KHJ. New Class of Materials: Half Metallic Ferromagnets. Phys Rev Lett 1983;**50**:2024.

[23] Fujii S, Sugimura S, Ishida S, Asano S. Hyperfine fields and electronic structures of the Heusler alloys Co_2MnX (X=Al,Ga,Si,Ge,Sn). J Phys Condens Matter 1990;**2**:8583.

[24] Singh D, Tiwari RS, Srivastava ON. Structural and magnetic properties of $Cu_{50}Mn_{25}Al_{25-x}Ga_x$ Heusler alloys. J Magn Mag Materials 2013;**328**:72.

[25] Nayak AK, Suresh KG, Nigam AK. Magnetic, electrical, and magnetothermal properties in Ni-Co-Mn-Sb Heusler alloys. J Appl Phys 2010;**107**:09A927.

[26] Luo H, Mang F, Jiang Q, Liu H, Liu E, Wu G, Wang Y. Effect of boron on the matensitic transformation and magnetic properties of $Ni_{50}Mn_{36.5}Sb_{13.5-x}B_x$ alloys. Scripta Materialia 2010;**63**:569.

[27] Chattopadhyay MK, Manekar MA, Sharma VK, Arora P, Tiwari P, Tiwari MK, Roy SB. Contrasting behavior of $Ni_{50}Mn_{35}In_{15}$ and $Ni_{50}Mn_{34.5}In_{15.5}$ alloys. J Appl Phys 2010;**108**: 073909

[28] Pathak AK, Khan M, Gautam BR, Stadler S, Dubenko I, Ali N. Exchange bias in bulk Ni- Mn-In based Heusler alloys. J Magn Mag Materials 2009;**321**:963.

[29] Gautam BR, Dubenko I, Pathak AK, Stadler S, Ali N. The structural and magnetic properties of $Ni_2Mn_{1-x}B_xGa$ Heusler alloys. J Magn Mag Materials 2009;**321**:29.

[30] Li S, Liu M, Yuan Z, Lu LY, Zhang Z, Lin Y, Du Y. Effect of Nb addition on the magnetic properties and magnetocaloric effect of CoMnSb alloy. J Alloys & Compds 2007;**427**:15.

[31] Singh S, Verma KC, Kotnala RK, Ralhan NK. Structure and magnetic properties of $NiFe_2O_4$nanoferrite prepared by chemical combustion route. AIP Conf Proc 2011;**1393**:215.

[32] McCurrie RA. Ferromagnetic Materials: Structure and Properties. Academic Press, New York, 1994.

[33] Hu W, Qin N, Wu G, Lin Y, Li S, Bao D. Opportunity of Spinel Ferrite Materials in Nonvolatile Memory Device Applications Based on Their Resistive Switching Performances. J Am Chem Soc 2012;**134**,14658.

[34] Zhu W, Wang L, Zhao R, Ren J, Lu G, Wang Y. Electromagnetic and microwave-absorbing properties of magnetic nickel ferrite nanocrystals. Nanoscale 2011;**3**:2862.

[35] Colombo M *et al.* Biological applications of magnetic nanoparticles. Chem Soc Rev 2012;**41**:4306.

[36] Perron H *et al*. Structural investigation and electronic properties of the nickel ferrite $NiFe_2O_4$: a periodic density functional theory approach. J Phys: Condens Matter 2007;**19**:346219.

[37] Verma KC, Kotnala R.K. Core/shell nanostructures due to Ce into $CoFe_2O_4$ induce lattice and vibrational defects for magnetic and dielectric enhancement. Adv Mater Lett 2016;**7(8)**:622.

[38] Szotek Z, Temmerman WM, Kodderitzsch D, Svane A, Petit L, Winter H. Electronic structures of normal and inverse spinel ferrites from first principles. Phys Rev B 2006;**74**: 174431.

[39] Luders U *et al*. Hybrid perovskite-spinel magnetic tunnel junctions based on conductive ferrimagnetic $NiFe_2O_4$. Appl Phys Lett 2006;**88**:082505.

[40] Singh S, Singh M, Ralhan NK, Kotnala RK, Verma KC. Improvement in ferromagnetism of $NiFe_2O_4$ nanoparticles with Zn doping. Adv Mat Lett 2012;**3(6)**:504.

[41] Singh S, Ralhan NK, Kotnala RK, Verma KC. Nanosize dependent electrical and magnetic properties of $NiFe_2O_4$ ferrite. Ind J Pure Appl Phys 2012;**50**:739.

[42] Singh D, Yadav TP, Tiwari RS, Srivastava ON. Phase formation in rapidly quenched Cu–based alloys. J Mater Sci. 2009;**44**:3883.

[43] Singh D, Yadav TP, Mandal RK, Tiwari RS, Srivastava ON. Effect of Ga substitution on the crystallization behaviour and glass forming ability of Zr-Al-Cu-Ni alloy. Mater Sci Eng A. 2010;**527**:469.

[44] Singh D, Mandal RK, Tiwari RS, Srivastava ON. Effect of cooling rate on the crystallization and mechanical behaviour of Zr-Ga-Cu-Ni metallic glass composition. J Alloys & Compds. 2015;**648**:456.

[45] Singh D, Mandal RK, Tiwari RS, Srivastava ON. Nanoindentation characteristics of $Zr_{69.5}Al_{7.5-x}Ga_xCu_{12}Ni_{11}$ glasses and their Nanocomposites. J Alloys & Compds 2011;**509**:8657.

[46] Singh D, Yadav TP, Mandal RK, Tiwari RS, Srivastava ON. Indentation characteristics of metallic glass and nanoquasicrystal–glass composite in Zr–Al(Ga)–Cu–Ni alloys. Intermetallics 2010;**18**:2445.

[47] Singh D, Singh M, Yadav TP, Mandal RK, Tiwari RS, Srivastava ON. Nanoindentation studies of metallic glasses and nanoquasicrystal-glass composites in Zr-Al (Ga)-Cu-Ni alloys. International Journal of Nanoscience 2011;**10**:929.

[48] Singh D, Yadav TP, Mandal RK, Tiwari RS, Srivastava ON. Effect of Ti addition on the quasicrystalline phase formation and indentation characteristics of $Zr_{69.5}Al_{7.5}Cu_{12}Ni_{11}$ alloy. Philosophical Magazine 2011;**91**:2837.

[49] Singh D, Singh D, Yadav TP, Mandal RK, Tiwari RS, Srivastava ON. Synthesis and indentation behavior of amorphous and nanocrystalline phases in rapidly quenched Cu–Ga–Mg–Ti and Cu–Al–Mg–Ti alloys. Metallography, Microstructure, and Analysis 2013;**2**:321.

[50] Bang D, Dan NH, Hung LT. Effect of microstructure on magnetism and transport properties of Cu-Mn-Al granular ribbons. J Phys D: Appl Phys 2008;**41**:155006.

[51] Sakka Y, Nakamura M. Mechanical and magnetic properties of the rapidly quenched Cu_2MnAl. J Mater Sci 1990;**25**:2549.

[52] Zhang M, Liu G, Cui Y, Hu H, Liu Z, Chen J, Wu G, Sui Y, Qian Z,

Zhang X. Magnetism and transport properties of melt-spun ribbon Cu_2MnAl Heusler alloy. J Magn Mag Materials 2004;**278**:328.

[53] Robinson J, McCormick P, Street R. Structure and properties of Cu_2MnAl synthesized by mechanical alloying. J Phys: Condens Matter 1995;7:4259.

[54] Heinzig M, Jenks CJ, Hove MV, Fisher I, Canfield P, Thiel PA. Surface preparation and characterization of the icosahedral Al–Ga–Pd–Mn quasicrystal. J Alloys & Compds 2002;**338**:248.

[55] Singh D, Singh D, Mandal RK, Srivastava ON, Tiwari RS. Glass forming ability, thermal stability and indentation characteristics of $Ce_{75}Al_{25-x}Ga_x$ metallic glasses. J Alloys & Compds 2014;**590**:15.

[56] Mandal RK, Tiwari RS, Singh D, Singh D. Influence of Ga substitution on the mechanical behavior of $Zr_{69.5}Al_{7.5-}xGa_xCu_{12}Ni_{11}$ and $Ce_{75}Al_{25-}xGa_x$ metallic glass compositions. MRS Proceeding. 2015;1757. doi:10.1557/opl.2015.45.

[57] Yadav TP, Singh D, Shahi RR, Shaz MA, Tiwari RS, Srivastava ON. Formation of quasicrystalline phase in $Al_{70-}xGa_xPd_{17}Mn_{13}$ alloys. Phil Mag. 2011;**91**:2474.

[58] Singh D, Singh D, Mandal RK, Srivastava ON, Tiwari RS. Effect of annealing on the devitrification behavior and mechanical properties of rapidly quenched Ce-based glassy alloys. J. Non-Cryst. Solids 2016;**445**:53.

[59] Singh D, Singh D, Mandal RK, Srivastava ON, Tiwari RS. Effect of quenching rate on the microstructure and mechanical behavior of $Ce_{75}Al_{21}Ga_4$ glassy alloy. Materials Characterization 2017;**134**:18.

[60] Konoplyuk SM, Kokorin VV, Kolomiets OV, Perekos AE, Nadutov VM. Magnetoresistance of Cu-Mn-Al melt-spun ribbons containing the system of interacting ferromagnetic inclusions. J Magn Mag Materials 2011;**323**:763.

[61] Lalla NP, Tiwari RS, Srivastava ON. Transmission electron microscopic investigations of rapidly solidified Al-Mn-Ni quasicrystalline alloys. Phil Mag B 1991;**63**:629.

[62] Verma KC, Singh VP, Ram M, Shah J, Kotnala RK. Structural, microstructural and magnetic properties of $NiFe_2O_4$, $CoFe_2O_4$ and $MnFe_2O_4$ nano ferrite thin films. J Magn Magn Mater 2011;**323**:3271.

[63] Singh S, Singh M, Ralhan N K, Kotnala RK, Verma KC. Ferromagnetic and Dielectric Properties of $Ni_{1-x}Zn_xFe_2O_4$ Nanoparticles Prepared via Chemical Combustion Route. Adv Sci Engg Med 2014; **6(6)**:688.

[64] Verma KC, Singh S, Tripathi SK, Kotnala RK. Multiferroic $Ni_{0.6}Zn_{0.4}Fe_2O_4$-$BaTiO_3$ nanostructures: Magnetoelectric coupling, dielectric, and fluorescence. J Appl Phys 2014;**116**:124103.

[65] Verma KC, Sharma AK, Bhatt SS, Kotnala RK, Negi NS. Synthesis and characterization of nanostructured $(Pb_{1-x}Sr_x)TiO_3$ thin films by a modified chemical route. Phil Mag 2009;**89(27)**:2321.

[66] Verma KC, Singh D, Kumar S, Kotnala RK. Multiferroic effects in $MFe_2O_4/BaTiO_3$ (M = Mn, Co, Ni, Zn) Nanocomposites. J Alloys Compds 2017;**709**:344.

[67] Sultan M, Singh R. FMR Studies on Nanocrystalline Zinc Ferrite Thin Films. J Phys: Conf Series 2010;**200**:072090.

[68] Verma KC, Kotnala RK. Nanostructural and lattice contributions to multiferroism in $NiFe_2O_4/BaTiO_3$. Mater Chem Phys 2016;**174**:120.

[69] Zhou JP, Lv L, Liu Q, Zhang YX, Liu P. Hydrothermal synthesis and

properties of $NiFe_2O_4$@$BaTiO_3$ composites with well-matched interface. Sci Technol Adv Mater 2012;**13**:045001.

[70] Wang P, Jin C, Zheng D, Li D, Gong J, Li P, Bai H, Strain and Ferroelectric- Field Effects Co-mediated Magnetism in (011)-$CoFe_2O_4$/ Pb($Mg_{1/3}Nb_{2/3})_{0.7}Ti_{0.3}O_3$ Multiferroic Heterostructures. ACS Appl Mater Interf 2016;**8**:24198.

[71] Bateer B *et al*. Synthesis, size and magnetic properties of controllable $MnFe_2O_4$ nanoparticles with versatile surface functionalities. Dalt Trans 2014;**43**:9885.

[72] Singh JP *et al*. Observation of bulk like magnetic ordering below the blocking temperature in nanosized zinc ferrite. J Magn & Mag Mater 2012;**324**: 2553.

[73] Goya GF, Rechenberg HR. Magnetic properties of $ZnFe_2O_4$ synthesized by ball milling. J Magn & Mag Mater 1999;**203**:141.

[74] Vamvakidis K *et al*. Reducing the inversion degree of $MnFe_2O_4$ nanoparticles through synthesis to enhance magnetization: evaluation of their 1 H NMR relaxation and heating efficiency. Dalton Trans 2014;**43**:12754.

[75] Bullita S *et al*. $ZnFe_2O_4$ nanoparticles dispersed in a highly porous silica aerogel matrix: a magnetic study. Phys Chem Chem Phys 2014;**16**:4843.

[76] Peddis D *et al*. Cationic distribution and spin canting in $CoFe_2O_4$ nanoparticles. J Phys Condens Matter 2011;**23**:426004.

[77] Carta D *et al*. A Structural and Magnetic Investigation of the Inversion Degree in Ferrite Nanocrystals MFe_2O_4 (M = Mn, Co, Ni). J Phys Chem C 2009;**113**:8606.

[78] Georgiadou V, Tangoulis V, Arvanitidid I, Kalogirou O, Samara CD. Unveiling the Physicochemical Features of $CoFe_2O_4$ Nanoparticles Synthesized via a Variant Hydrothermal Method: NMR Relaxometric Properties. J Phys Chem C2015;**119**:8336.

[79] Smith S, Wijn HPJ. Ferrites, Philips Library, Amsterdam, 1961

[80] Rezlescu N, Rezlescu E, Pasnicu C, Craus ML. Effects of the rare-earth ions on some properties of a nickel-zinc ferrite. J Phys: Condens Matter 1994;**6**:5707.

[81] Sun QC, Birkel CS, Cao J, Tremel W, Musfeldt JL. Spectroscopic Signature of the Superparamagnetic Transition and Surface Spin Disorder in $CoFe_2O_4$ Nanoparticles.ACS Nano 2012;**6**:4876.

[82] Kotnala RK, Shah J, Mathpal MC, Verma KC, Singh S, Lovkush. Influence of annealing on humidity response of RF sputtered nanocrystalline $MgFe_2O_4$ thin films. Thin Solid Films 2011;**519**:6135.

Chapter 6

Effect of M Substitution on Structural, Magnetic and Magnetocaloric Properties of R_2Fe_{17-x} M_x (R = Gd, Nd; M = Co, Cu) Solid Solutions

Mosbah Jemmali and Lotfi Bessais

Abstract

The structure, magnetic and magnetocaloric properties of $Nd_2Fe_{17-x}Co_x$ (x = 0; 1; 2; 3, 4) and $Gd_2Fe_{17-x}Cu_x$ (x = 0, 0.5, 1 and 1.5) solid solutions have been studied. For this purpose, these samples were prepared by arc melting and subsequent annealing at 1073 K for a 7 days. Structural analysis by Rietveld method on X-ray diffraction (XRD) have determined that these alloys crystallize in the rhombohedral Th_2Zn_{17}-type structure (Space group R¯3 m) and the substitution of iron by nickel and copper leads to a decrease in the unit cell volume. The Curie temperature (T_C) of the prepared samples depends on the nickel and copper content. Based on the Arrott plot, these analyses show that $Nd_2Fe_{17-x}Co_x$ exhibits a second-order ferromagnetic to paramagnetic phase transition around the Curie temperature. These curves were also used to determine the magnetic entropy change ΔS_{Max} and the relative cooling power. For an applied field of 1.5 T, ΔS_{Max} increase from 3.35 J/kg. K for x = 0 to 5.83 J/kg. K for x = 2. In addition the RCP increases monotonously. This is due to an important temperature range for the magnetic phase transition, contributing to a large ΔS_{Max} shape. $Gd_2Fe_{17-x}Cu_x$ solid solution has a reduction of the ferromagnetic phase transition temperature from 475 K (for x = 0) to 460 K (for x = 1.5) is due to the substitution of the magnetic element (Fe) by non-magnetic atoms (Cu). The magnetocaloric effect was determined in the vicinity of the Curie temperature T_C. By increasing the Cu content, an increase in the values of magnetic entropy (ΔS_{Max}) in a low applied field is observed.

Keywords: Rare-earth alloys and compounds, magnetization, magnetocaloric effect

1. Introduction

During the last decades and until now, the production of cold has mainly been ensured by the technique of compression/expansion of a refrigerant. This process is developed and reliable, however it has a large number of disadvantages due to the use of toxic gases such as chlorofluorocarbon (CFC) or hydrochlorofluorocarbon (HCFC) which have proved to be very harmful to the environment (destruction of the ozone layers) and contribute to the greenhouse effect. Current environmental requirements and ecological standards limit conventional technologies. It is for this reason that researchers and manufacturers of refrigeration and heat pump have set out to search for a new refrigeration technology that is more respectful of the environment and less energy intensive, which is magnetic refrigeration (RM). This cold production technology, which is based on a physical phenomenon called the magnetocaloric effect (EMC), has considerable advantages over conventional techniques: absence of atmospheric pollutants, absence of noise and vibration, high reliability thanks to the use solid refrigerants rather than harmful gases and above all environmental protection and reduction of electricity consumption.

In this regard, in order to get rid of harmful refrigerants, additional efforts have been turned towards the search for magnetocaloric materials. Much attention in this area has been focused on intermetallic compounds which are defined as solid phases containing two or more metals, possibly with one or more non-metallic elements, whose crystal structure differs from that of the constituent elements. In 1997, Pecharsky and Gschneidner reported the discovery of a giant magnetocaloric effect in the intermetallic compounds Gd_5 $(Si_xGe_{1-x})_4$ and $Gd_5(Si_xGe_2)$ [1–3] which are currently the benchmarks in the field of magnetic refrigeration.

Intermetallic compounds combining rare earths and transition metals have been the subject of much research work in recent years, aiming at their development for technological applications such as magnetic refrigeration, aeronautical turbines [4], battery electrodes [5], and the development of high performance permanent magnets [6] such as the compound $Nd_2Fe_{14}B$ which has a coercive field equal to 2.3 T [7]. These magnetic materials are now model compounds combining the localized magnetism of rare earth elements and the less localized, or even itinerant, of transition elements.

Long-range magnetic order is mainly found in intermetallic compounds that are based on 4f rare-earth metals (R) and 3d transition metals (M), and in which occur three types of exchange interactions, namely: (1) the 3d–3d exchange interaction (JMM) between the magnetic moments of the M sublattice, (2) the 4f–4f exchange interaction (JRR) between the magnetic moments within the R sublattice, and (3) the intersublattice 3d–4f exchange interaction (JRM). It is noteworthy that the interactions between the rare-earth spins 4f–4f are supposed to be negligible in comparison with the other two types of interactions [8–19].

Recent interest in these R_2Fe_{17} intermetallic compounds has been renewed due to the magnetocaloric properties shown by these ferromagnetic compounds Pr_2Fe_{17}, Nd_2Fe_{17}, Er_2Fe_{17}, Gd_2Fe_{17}, Sm_2Fe_{17} and Tb_2Fe_{17} [20–25]. The selected ferromagnetic compound Gd_2Fe_{17} studied in this research has a magnetocaloric effect with a maximum magnetic entropy ΔS_{Max} equal to 0.89 J/kg K, for an external field change from 0 T to 1.5 T at room temperature [24]. The goal here was to study the M effect on structural, and magnetic properties of $R_2Fe_{17-x}Co_x$. In our case the Fe and M structure factors are very similar, which makes difficult to localize the M atom sin the structure using the X-ray powder diffraction.. Finally, we have studied the low field magnetic entropy changes in R_2Fe_{17} compounds and we have demonstrated the magnetocaloric effect enhancement due to M substitution, observed for the first time for intermetallic compounds. A systematic study of the constitutional properties of a ternary phase diagram formed by Sm-Fe-Ni and Nd-Fe-Co have been studied by our group [26, 27]. The substitution of Ni or Co by Fe atoms confirmed a extension of binaries Nd_2Fe_{17} and $GdFe_{17}$ in the Nd-Fe-Co and Nd-Fe-Co ternary systems, respectively and improved physical support properties and structural stability [28, 29].

2. Crystallographic study of the solid solution $R_2Fe_{17-x}M_x$

The R_2Fe_{17} binaries are generally either of rhombohedral symmetry of type Th_2Zn_{17} ($R\bar{3}m$) for light rare earths (from Ce to Gd), or of hexagonal symmetry of type Th_2Ni_{17} ($P6_3/mmc$) for heavy rare earths, such as Tb and Dy. These structures derive from the $CaCu_5$ structure and differ in the stacking mode of the $CaCu_5$ unit entities in **Figures 1** and **2**.

- Th_2Zn_{17} has nine entities per cell where three rare earth atoms have each been replaced by an M-M dumbbell (Dummbell site). This description can be schematized by:

$$9\ RM_5 - 3\ R + 3\ x\ (2M) = 3\ R_2M_{17} \qquad (1)$$

- Th_2Ni_{17} has six entities per cell where two rare earth atoms have each been replaced by a M. M dumbbell (Dummbell site). This description can be schematized by:

$$6\ RM_5 - 2\ R + 2\ x\ (2M) = 2\ R_2M_{17} \qquad (2)$$

The equilibrium binary compound Nd_2Fe_{17} crystallizes in the Th_2Zn_{17} structure type of space group $R\bar{3}m$ in **Figure 2** with the lattice parameters a = 8.5796 (3) Å and

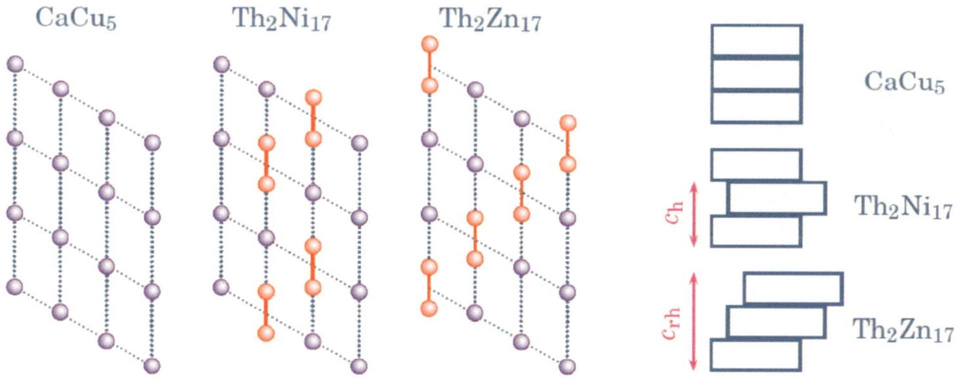

Figure 1.
2/17 stoichiometric structures derived from $CaCu_5$: Th_2Zn_{17} ($R\bar{3}m$) and Th_2Ni_{17} ($P6_3/mmc$).

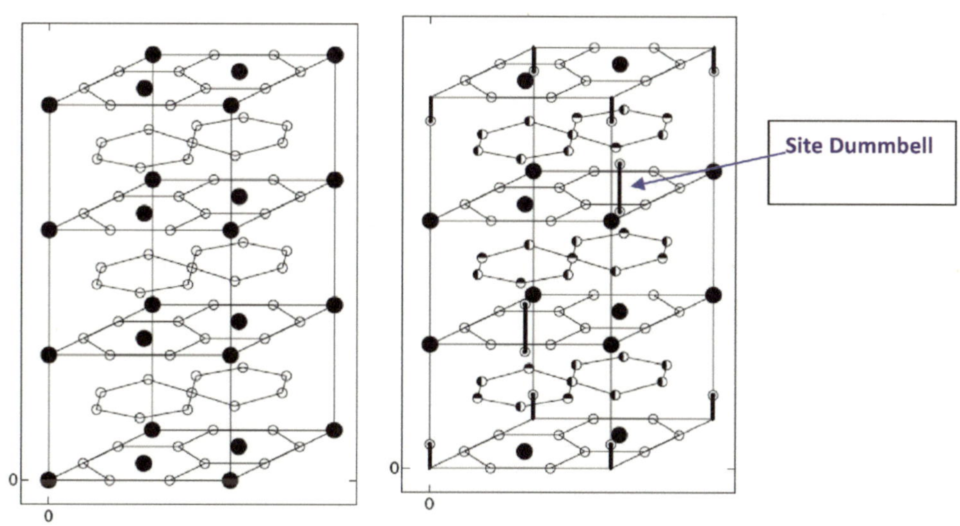

Figure 2.
Mailles P6/mmm de type $CaCu_5$(a) et $R\bar{3}m$ de type Th_2Zn_{17} (b).

c = 12.4624 (2) Å. They agree with those found in the literature [30]. This rhombohedral phase $R\bar{3}m$ can be stabilized by certain elements such as Co [31], Cr [32], Si [33] or Ga [34]. The compound Sm_2Fe_{17} as representative of the solid solution crystallizes in a rhombohedral cell of space group $R\bar{3}m$ as was mentioned by X. C. Kou et al. [35]. The atoms of the rare earth occupy the crystallographic sites 6c while the iron atoms occupy 4 unequal sites: 6c, 18 h, 18f and 9d.

The most widely used synthesis method for the production of intermetallic is the melting method, which makes it possible to obtain materials in a solid state followed by annealing or grinding after annealing. Using arc furnace melting followed by annealing for one week, well adequate to ensure good crystallization and complete atomic diffusion. To control the stoichiometry and to avoid impurities, we have optimized the conditions for developing these phases using this method from work published by our laboratory [36, 37]. The development step is followed by a crystallographic study which allowed us to determine the nature of the site occupied by the metal transition and metalloids as well as the crystallographic parameters. This structural study is done by the X-ray diffraction method followed by Rietveld refinement, coupled with studies by scanning electron microscopy.

3. Structural of the solid solution $Nd_2Fe_{17-x}Co_x$ ($0 \leq x \leq 4$)

The samples synthesized along the $Nd_2Fe_{17-x}Co_x$ line show the existence of a solid solution which crystallizes in an $R\bar{3}m$ cell of Th_2Zn_{17} type structure extending along the $0 \leq x \leq 4$ domain.

The lattice parameters and atomic positions determined from the refinement of the few compositions synthesized on the line of the binary extension $Nd_2Fe_{17-x}Co_x$ ($0 \leq x \leq 4$) are grouped together in **Tables 1** and **2**.

The Rietveld refinements of the $Nd_2Fe_{17-x}Co_x$ series ($0 \leq x \leq 4$) are shown in **Figure 3**.

Compositions	*a* (Å)	*c*(Å)	*c/a*	$\chi 2$	R_B
x = 0	8.5792(2)	12.4615(2)	1.452	1.97	7.18
x = 1	8.5763(2)	12.4606(3)	1.452	3.57	6.78
x = 2	8.5597(2)	12.5069(4)	1.461	2.47	3.3
x = 3	8.5594(3)	12.5081(4)	1.463	2.77	3.3
x = 4	8.5598(3)	12.5069(4)	1.461	2.47	7.51

Table 1.
The lattice parameters of the $Nd_2Fe_{17-x}Co_x$ system ($0 \leq x \leq 4$).

Atomes	Positions de Wyckoff	x	y	z	Occupation
Nd	*6c*	0.00000	0.00000	0.34304	1
Fe1	*18f*	0.29276	0.00000	0.00000	1
Fe2	*9d*	0.50000	0.00000	0.50000	1
Fe3	*6c*	0.00000	0.00000	0.09938	1
Fe4	*18 h*	0,5	0,5	0.15679	0,334
Co4	*18 h*	0,5	0,5	0.15679	0,666

Table 2.
Atomic positions (x, y, z) and occupation of the various Wyckoff positions for the compound $Nd_2Fe_{13}Co_4$.

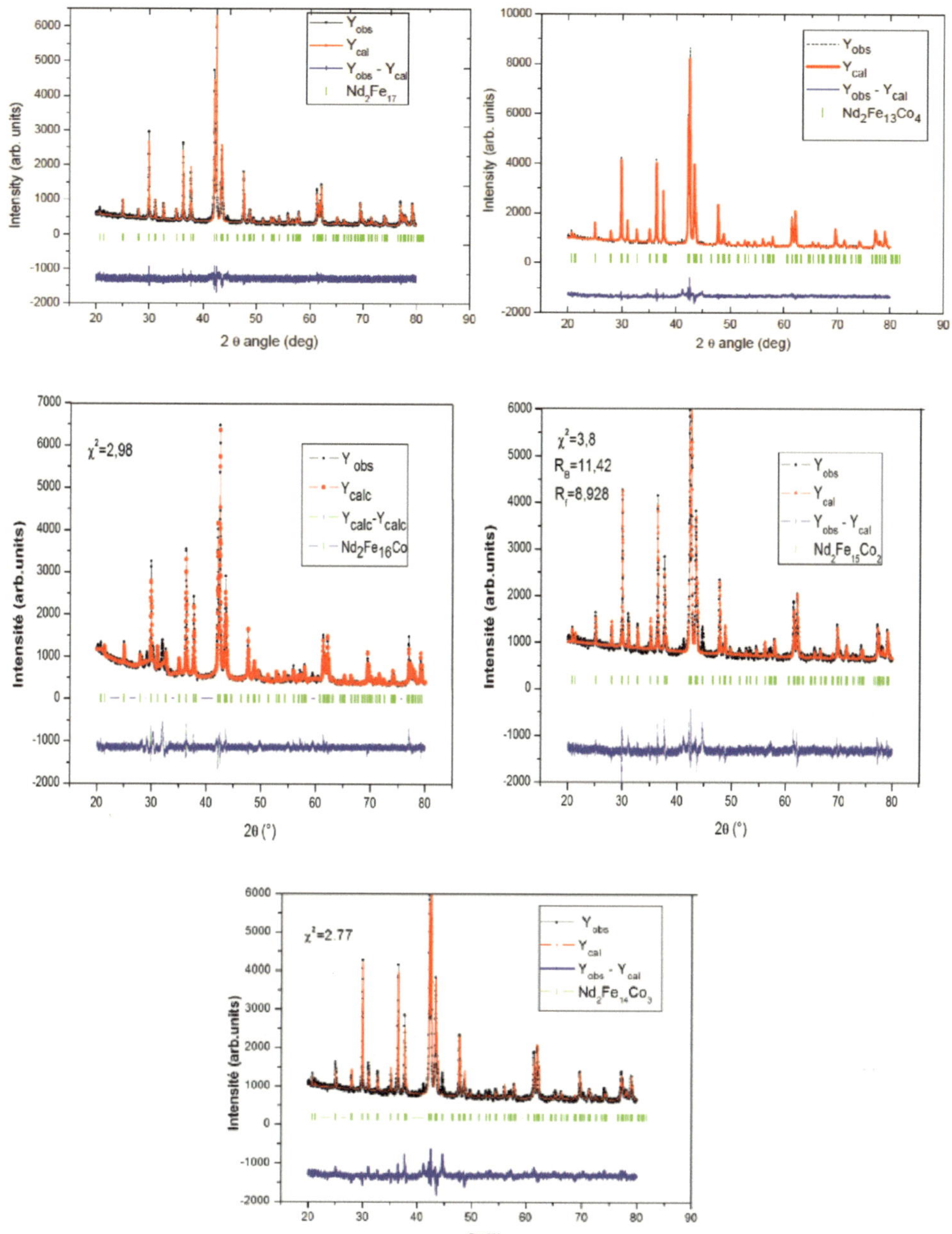

Figure 3.
Rietveld refinements results of the solid solution $Nd_2Fe_{17-x}Co_x$ ($0 \leq x \leq 4$).

The unit cell parameters for Nd_2Fe_{17} obtained are: a = 8.5792(2) Å, c = 12.4615(2) Å, and a = 8.5598(3) Å, c = 8.5598(3) Å, for Nd_2Fe_{17} and $Nd_2Fe_{13}Co_4$ compounds, respectively.

The effect of Co on the lattice parameter is quite weak up to x = 4, the parameter a shows a tendency to decrease, $\Delta a/a \approx -5.6$ x 10-4 per atom of cobalt, on the other hand the parameter c increases slightly $\Delta c/c = 9.3$ x 10-4 per atom of cobalt, but the volume and the c/a ratio remains almost constant (**Figure 4**). While c/a is almost constant. The atomic positions are not affected by the substitution.

Our results are in agreement with those of Li et al. [38]. For these authors c increases until x = 2 whereas we have measured without any ambiguity a slight

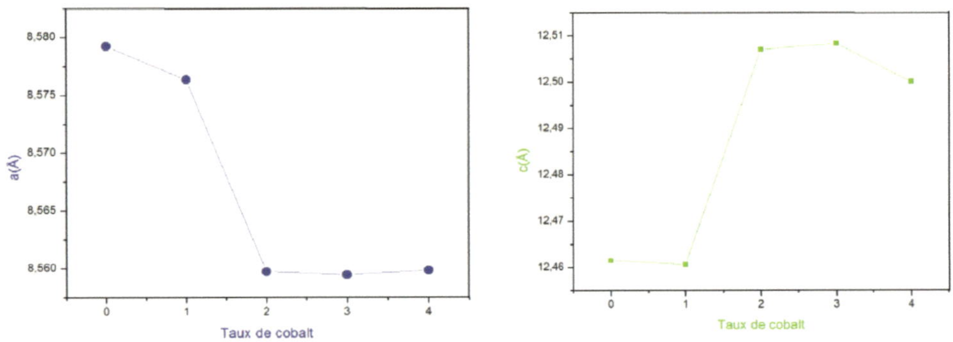

Figure 4.
Variation of parameters a and c for $Nd_2Fe_{17-x}Co_x$, as a function of the level of cobalt x.

increase. On the other hand, our results are in agreement with the decrease found by Lin et al. [39]. Atomic positions are not affected by substitution. A. Nandra et al. [40], obtained the converse. There is also a decrease in c, equal to $6.5x10^{-3}$Å per small silicon atom compared to the variation in an equal to $19x10^{-3}$Å per silicon atom. Moreover, the parameter values found for $Sm_2Fe_{15}Si_2$ and $Sm_2Fe_{16}Si$ [41, 42], confirm these results. In general, the evolution of lattice parameters, in solid solutions $R_2Fe_{17-x}Si_x$, depends on the nature of the rare earth R. For R = Ce, a decreases while c increases, for R = Dy, Y both parameters decrease [43, 44]. A simple steric effect of substitution of iron with radius r_{Fe} = 1.274 Å by a smaller atom can nevertheless explain such an evolution if we consider the covalent radius of cobalt equal to 1.252 Å.

4. Structural of the solid solution: $Gd_2Fe_{17-x}Cu_x$ $0 \leq x \leq 1.5$

Scanning electron microscopy analyzes coupled with X-ray diffractograms identified using PowderCell software reveal a solubility domain of $Gd_2Fe_{17-x}Cu_x$ that extends up to x = 1.5. Whereas, the solubilities of Ti, Mo and Re in the binary Gd_2Fe_{17} [39–41] are of order 2.3; 0.06 and 1.5 at%, respectively. The X-ray powder diffraction diagram (**Figure 5**) of the nominal composition Gd6-Fe14-Cu9 is

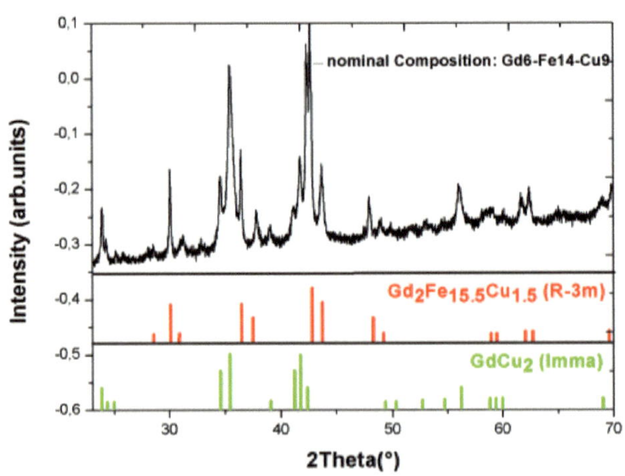

Figure 5.
X-ray diffractogram of the alloy of nominal composition Gd6-Fe14-Cu9 located in binary region 4 showing the bi-phasic equilibrium between: $GdCu_2$ + $Gd_2Fe_{15,5}Cu_{1,5}$.

indexed on the basis of the two rhombohedral and orthorhombic cells indicating the equilibrium thermodynamics between the compound $GdCu_2$ and the limit of the solid solution $Gd_2Fe_{17-x}Cu_x$ with x = 1.5 [45–47].

The SEM image reconstituted from backscattered electrons of the compound $Gd_6Fe_{14}Cu_9$ annealed at 800° C (**Figure 6**) is in good agreement with the result found by the X-ray powder diffraction, since we see there two main types of contrasts which correspond to each of the two phases $GdCu_2$ and $Gd_2Fe_{15.5}Cu_{1.5}$.

To show the extension of the binary Gd_2Fe_{17} in the ternary system, and to make a study of the structural, we synthesized three single-phase samples along the line $Gd_2Fe_{17-x}Cu_x$ (x = 0.5; 1 and 1.5). **Figure 7** shows the Rietveld refinement of the X-ray diffractograms of those compounds which crystallize in the rhombohedral structure with the R-$\bar{3}$m space group. These diffractograms show no structural phase transition following the substitution of iron by copper.

We have demonstrated the formation of a new solid solution $Gd_2Fe_{17-x}Cu_x$ ($0 \le x \le 1.5$). Following the Rietveld refinement, it was concluded that the limit of

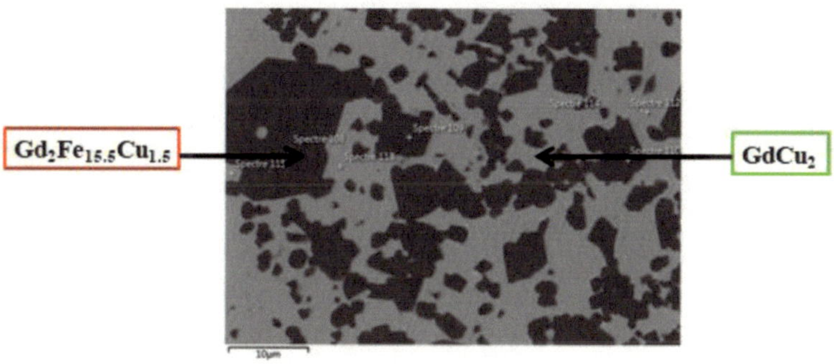

Figure 6.
SEM-EDS image of the alloy of nominal composition Gd6-Fe14-Cu9 showing the limit of solid solution $Gd_2Fe_{17-x}Cu_x$ and the three-phase equilibrium between $GdCu_2$ and $Gd_2Fe_{15.5}Cu_{1.5}$.

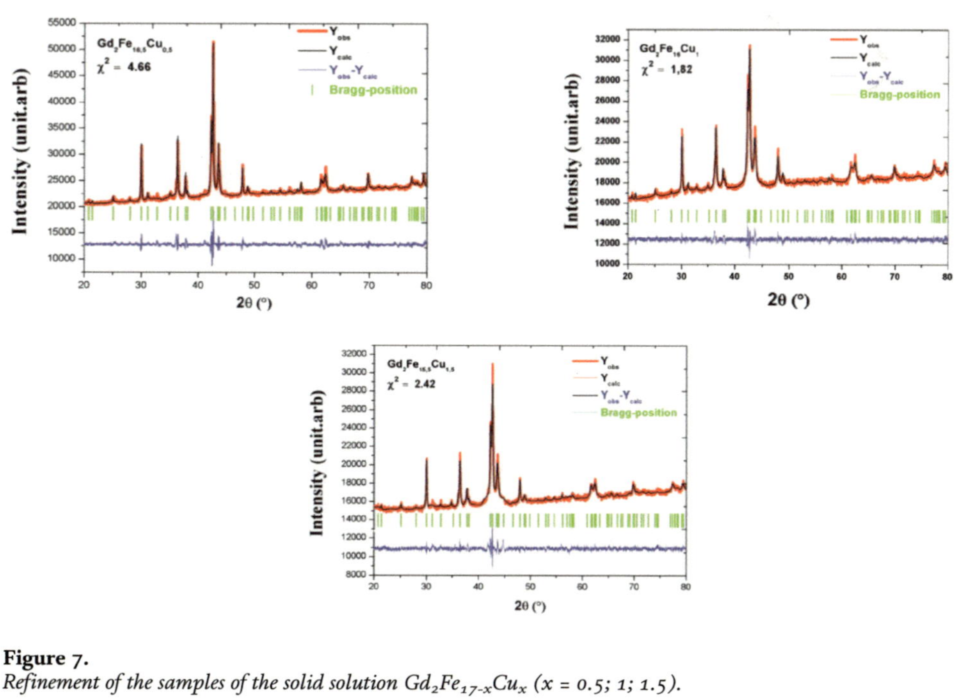

Figure 7.
Refinement of the samples of the solid solution $Gd_2Fe_{17-x}Cu_x$ (x = 0.5; 1; 1.5).

the solid solution $Gd_2Fe_{17-x}Cu_x$ corresponds to $Gd_2Fe_{15.5}Cu_{1.5}$. The results of structural analyzes obtained by the Rietveld refinement of X-ray diffraction data are referenced in **Table 3**.

The Rietveld refinements of X-ray diffraction diagrams made it possible to follow the evolution of lattice parameters as a function of the copper content. The decrease in the lattice parameters a and c in these compounds following the substitution of iron by copper can be explained in terms of the atomic size of the element substituted for Fe. We also notice that the lattice volume of the $Gd_2Fe_{17-x}Cu_x$ system decreases by increasing the Cu level.

The Rietveld refinement of the compounds of the solid solution $Gd_2Fe_{17-x}Cu_x$ was carried out according to the standard procedure. For an occupation of copper at site 6c, 9d, 18 h or 18f, the lattice parameters, atomic positions and line profile parameters were considered as adjustable parameters. The best agreement is found for a copper substitution at the 18 h site. From this result, we concluded that copper only substitutes at the 18 h site. However, the Rietveld refinement of $Gd_2Fe_{17-x}Ti_x$ studied by G. Pokharel et al. [48] proves that the substitution of titanium by iron was made in the two sites 18 h and 18 f. For the solid solution $Gd_2Fe_{17-x}Si_x$, the preferred silicon substitution site is 18 h [49].

Figure 8 shows the rhombohedral crystal lattice of the compound $Gd_2Fe_{16.5}Cu_{0.5}$. The gadolinium atoms occupy the crystallographic site (6c) with an occupancy rate equal to 1 and the atoms Fe1, Fe2, and Fe3 are located respectively in the three unequal sites 6c, 9d, 18f with an occupancy rate of order 1. The two atoms

Composition	Groupe d'espace	a (Å)	c (Å)	V (Å³)	χ^2	R_B	R_F
$Gd_2Fe_{16,5}Cu_{0,5}$	$R\bar{3}m$	8,539(6)	12,436(7)	785,554(5)	4,66	5,37	5,10
$Gd_2Fe_{16}Cu_1$	$R\bar{3}m$	8,534(4)	12,433(3)	784,254(2)	1,82	3,54	3,57
$Gd_2Fe_{15,5}Cu_{1,5}$	$R\bar{3}m$	8,533(5)	12,426(6)	783,748(4)	2,42	11,2	15,3

Table 3.
Crystallographic parameters of the solid solution $Gd_2Fe_{17-x}Cu_x$ (x = 0.5; 1 and 1.5).

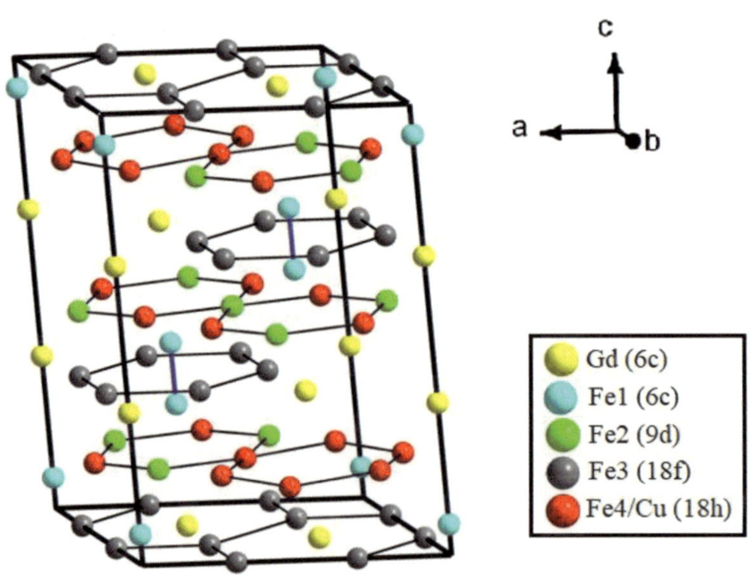

Figure 8.
Schematic representation of a rhombohedral $R\bar{3}m$ crystal structure of $Gd_2Fe_{16.5}Cu_{0.5}$.

Atomes	Sites de Wyckoff	Position des sites			Occupation
		X	Y	Z	
Gd	6c	0	0	0,336	1
Fe1	6c	0	0	0,075	1
Fe2	9d	1/2	0	1/2	1
Fe3	18f	0,303	0	0	1
Fe4	18 h	0,501	0,499	0,160	0.916
Cu	18 h	0,501	0,499	0,160	0.084

Table 4.
Characterization of the atomic sites of the compound $Gd_2Fe_{16.5}Cu_{0.5}$.

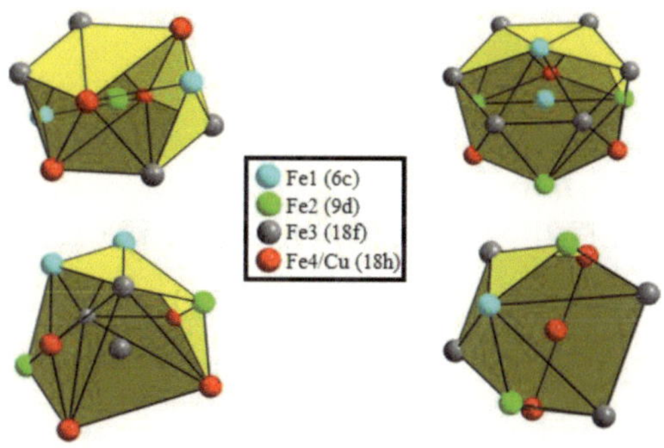

Figure 9.
The coordination polyhedra for each crystallographic Fe site.

Fe4 and Cu are distributed over the same Wyckoff site (18 h) with an occupancy rate equal to 0.916 and 0.084 respectively. The atomic positions for the different crystallographic sites of the R-$\bar{3}$m structure are reported in **Table 4**.

In this type of structure we notice the presence of layers formed by hexagons. The hexagons made up of Fe3 atoms (18f) contain gadolinium atoms at the centers. While the hexagons formed by the Fe2 atoms located in the sites (9d) and the Fe4/Cu atoms located in the sites (18 h) are empty.

Figure 9 shows the coordination polyhedra of the four sites: 6c, 9d, 18f and 18 h. The polyhedron of the Fe2 atom located in site (9d) is an icosahedron consisting of four atoms of Fe4 / Cu (18 h), four atoms of Fe3 (18f) and two atoms of Fe1 (6c). In addition, the Fe1 atom located in site (6c) is surrounded by six Fe3 atoms (18f), three Fe4 / Cu atoms (18 h), three Fe2 atoms (9d) and a single Fe1 atom (6c). Also, the Fe3 atom located in site (18f) is coordinated by two Fe2 atoms (9d), two Fe3 atoms (18f), four Fe4 / Cu atoms (18 h) and two Fe1 atoms (6c) thus forming an icosahedron. Finally, the polyhedron of the Fe3 atom located in site (18 h) is made up of two atoms of Fe2 (9d), four atoms of Fe3 (18f), a single atom of Fe1 (6c) and two atoms of Fe4/Cu (18 h).

5. Magnetic and manetocaloric properties of the solide solution $R_2Fe_{17-x}M_x$

The Curie temperature is the temperature of the transition from the ordered magnetic state (ferromagnetic) to the disordered state (paramagnetic).

This transition is manifested by a sharp drop in magnetic susceptibility measured as a function of temperature.

Remember that in R-Fe intermetallics, the order temperature is determined by the interactions between iron atoms. The nature of these interactions depends on the degree of filling of the 3d band of iron and the Fe-Fe distances. Interatomic distances of less than 2.45 Å promote antiferromagnetic interactions [50]. The Sm_2Fe_{17} compound is characterized by a low Curie temperature [51]. This is mainly due to the short inter-atomic Fe-Fe distances of 6c-6c dumbbells and as well as 9d-18f distances, the corresponding distances being 2.39 Å and 2.44 Å respectively. These distances, less than 2.45 Å, lead to negative Fe-Fe interactions, because these atoms are anti-ferromagnetically coupled [52].

In general, in rare earth intermetallic compounds, the Curie temperature is given by three types of exchange interactions:

- The 3d-3d ($J_{Fe\text{-}Fe}$) exchange between the magnetic moments of the sub-network of iron atoms.

- The 4f -4f exchange ($J_{R\text{-}R}$) between the magnetic moments of rare earth atoms.

- The 3d -4f ($J_{R\text{-}Fe}$) exchange between the two 3d-4f subnets.

Of these three types of exchange interactions, the 4f -4f exchange is the weakest and can be overlooked. However, if we consider an R-Fe system where R is non-magnetic, we can also neglect the interactions between the two sub-networks (3d-4f) and we can consider that the only contribution to the Curie temperature is due to the interaction between the magnetic moments of the iron sub-lattice (3d-3d).

The crystal structure dictates the mode of interaction as well as the intensity of the interaction. These interactions seem to originate from the electronic coupling between close neigh boring atoms. The intensity of these interactions is a factor in the distance between carriers. The curve in **Figure 10** shows the exchange

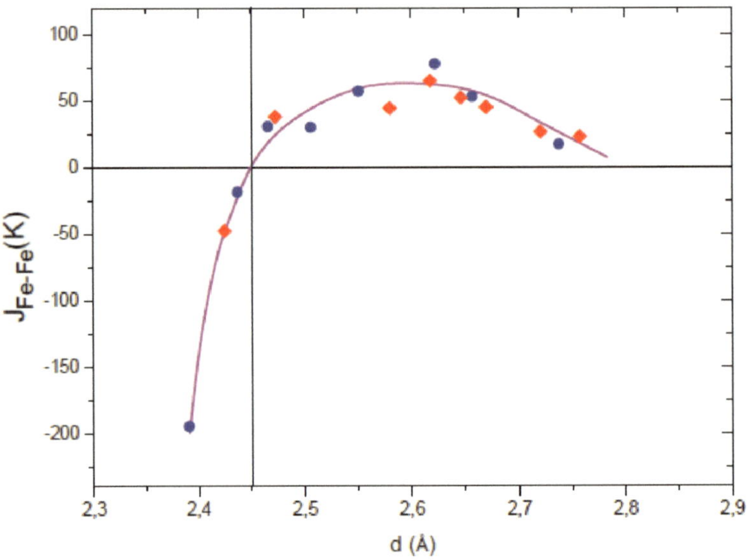

Figure 10.
Exchange interaction as a function of iron–iron distances for the Sm_2Fe_{17} compounds (in lozange) and their nitrides (in circle).

interaction as a function of iron–iron distances for the (Gd, Nd)$_2$Fe$_{17}$ compounds and their nitrides. The signs and intensity of this integral are closely related to the distance between Fe-Fe for both phase 2/17 and its nitride. Indeed for interatomic distances less than 2.45 Å leads to negative Fe-Fe interactions where the atoms are coupled antiferromagnetically. Beyond that, these interactions become positive, which corresponds to a ferromagnetic coupling between close neighboring iron. The exchange interaction is very intense, but only acts between nearby moments and subsides very quickly with distance. The Curie temperature is the result of two effects: a magnetovolumic effect [53, 54] linked to Fe-Fe distances and an electronic effect linked to the filling of the 3d band of iron.

In order to obtain as much information as possible on the magnetocaloric effect in our intermetallics in the vicinity of their Curie temperature. The variation of the magnetic entropy ΔS_{Max} is evaluated by an indirect method whose main ingredient is magnetic measurement; this method is therefore based on magnetization isotherms as a function of the magnetic field applied for different temperatures. This amounts, for each temperature, to calculating the area between the two isotherms around the temperature of Tc using the following equations:

$$\Delta S_M(T, \mu_O H) = S_M(T, \mu_0 H) - S_M(T, 0) = \int_0^{\mu_0 H_{\max}} \left(\frac{\partial S}{\partial(\mu_0 H)}\right) d(\mu_0 H) \tag{3}$$

From Maxwell's thermodynamic relation:

$$\left(\frac{\partial S}{\partial(\mu_0 H)}\right)_T = \left(\frac{\partial M}{\partial T}\right)_{\mu_0 H} \tag{4}$$

We can get the equation:

$$\Delta S_M(T, \mu_O H) = S_M(T, \mu_0 H) - S_M(T, 0) = \int_0^{\mu_0 H_{\max}} \left(\frac{\partial M}{\partial T}\right)_{\mu_0 H} d(\mu_0 H) \tag{5}$$

Finally, the magnetic entropy was calculated, using software, by the expression:

$$\Delta S_M\left(\frac{T_1 - T_2}{2}\right) = \left(\frac{1}{T_1 - T_2}\right)\left[\int_0^{\mu_0 H_{\max}} M(T_2, \mu_0 H) d(\mu_0 H) - \int_0^{\mu_0 H_{\max}} M(T_{1,}\mu_0 H) d(\mu_0 H)\right] \tag{6}$$

6. Magnetic and magnetocaloric properties of the solid solution $Nd_2Fe_{17-x}Co_x$

6.1 Magnetic properties

Magnetic study of the binary compound Nd_2Fe_{17} reveals the existence of a Curie temperature of the order of T_C = 326 K. These samples show a ferro-paramagnetic type transition. The Curie T_C temperature was determined from the curves (dM/dT). This temperature increases as the concentration of cobalt increases while indicating an increase in ferromagnetism from **Figure 11**. Indeed, Co plays a main role in strengthening 3d-3d interactions. Based on the very short interatomic distances, the low Curie temperature of the Nd_2Fe_{17} compound is due to the level of the Fe-Fe dumbbell pairs, located at the 6c site where the iron atoms are antiferromagnetically coupled.

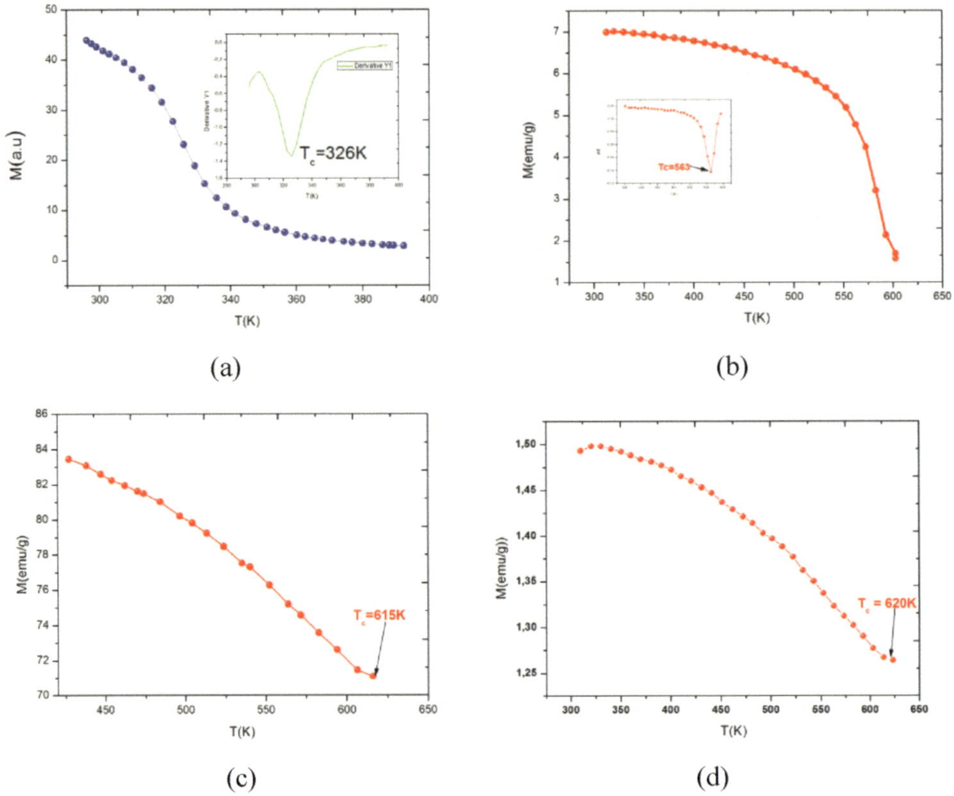

Figure 11.
Magnetization as a function of temperature M (T) for Nd_2Fe_{17} (a) $Nd_2Fe_{15}Co_2$ (b) $Nd_2Fe_{14}Co_3$ (c) $Nd_2Fe_{13}Co_4$ (d).

The increase in Curie temperature with the level of cobalt can be attributed, in part, to a reduction in the number of Fe-Fe pairs, coupled antiferromagnetically. Substitution by cobalt induces filling of the 3d band of iron, which promotes positive 3d-3d interactions [55]. In addition, the effect of cobalt is to slightly shift 3d states to higher energies. This more pronounced effect in the minority spin band leads to an increase in iron moment and is partially responsible for the increase in Curie temperature.

6.2 Magnetocaloric effect

The variation of the magnetic entropy ΔS_{Max} is evaluated by an indirect method whose main objective is the magnetic measurement; this method is based on isotherms magnetization depending on the magnetic field applied for different temperatures. Indeed, for each temperature, we calculate the area between the two isotherms around the temperature chosen in a range of field divided by the temperature difference between these two curves.

The calculated ΔS value for each applied field was calculated from the curves of the isotherm M (H) (**Figure 12**) using the following equations (Maxwell's relation) Eqs. (1)–(4):

Dans le but de connaitre la nature de la transition ferro-paramagnétique, nous avons porté sur la **Figure 13** les isothermes d'Arrott [56] donnant M^2 en fonction de H/M pour les différentes températures pour toute la gamme de composition x en cobalt.

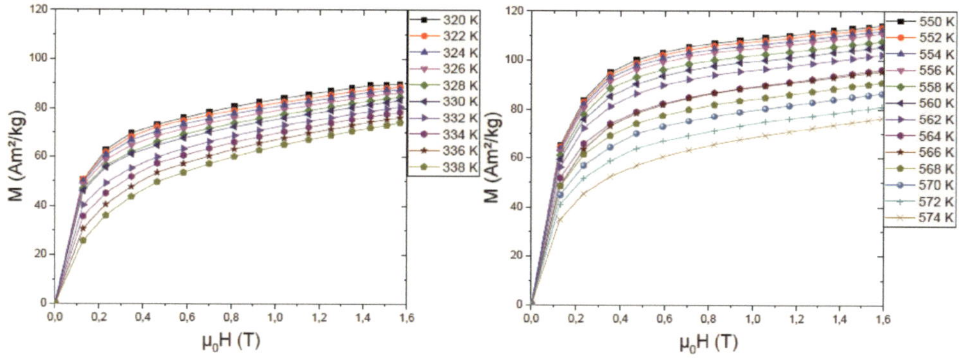

Figure 12.
Variation of the magnetization as a function of the magnetic field applied for the compounds of Nd_2Fe_{17} (left) and $Nd_2Fe_{15}Co_2$ (right).

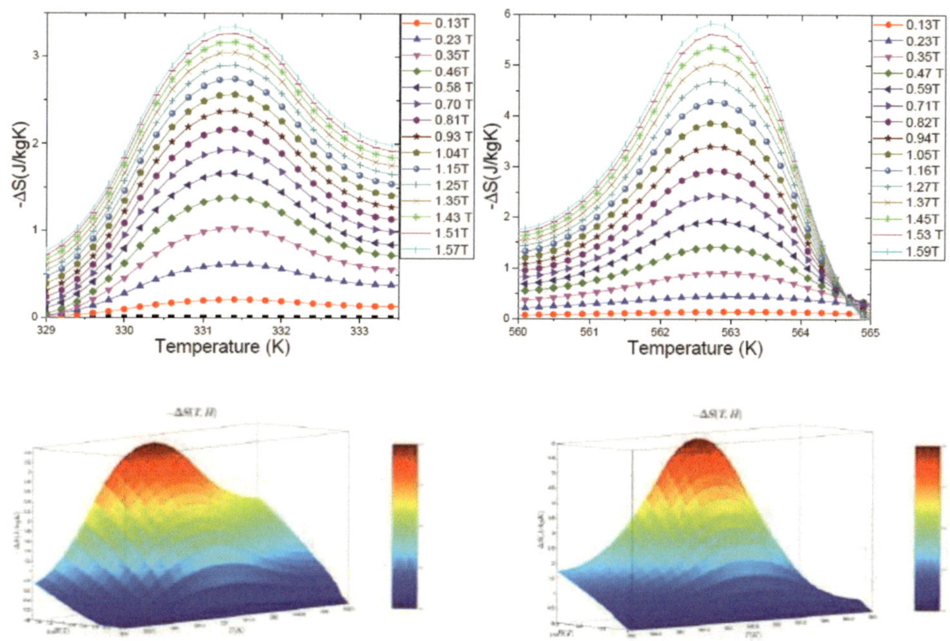

Figure 13.
Variation of magnetic entropy $-\Delta S$ (T) depending on the temperature and the magnetic field of Nd_2Fe_{17} (left) and $Nd_2Fe_{15}Co_2$ (right).

The $Nd_2Fe_{17-x}Co_x$ magnetocaloric properties were determined and calculated from the magnetization measured as a function of the magnetic field at different constant temperatures. In order to continue the study of the effect of cobalt on the magnetic and magnetocaloric properties of our compound, we studied the change of entropy $-\Delta S$ (T). The variation of the magnetic entropy ΔSM is evaluated by an indirect method whose the main objective is magnetic measurement, this method is based on magnetization isotherms as a function of the magnetic field applied for different temperatures. Indeed, for each temperature, we calculate the area between the two isotherms around the temperature chosen in a range of field divided by the temperature difference between these two curves.

For magnetocaloric applications, it is interesting to determine the relative refrigeration capacity (which we will note RCP or relative cooling power). It is the

x	0	2
Tc	331	563
($-\Delta S_{Max}$) (J/kg.K)	3.35	5.83
RCP(J/kg)	11.6	16

Table 5.
Summary of magnetocaloric properties of $Nd_2Fe_{17-x}Co_x$ system (x = 0 and x = 2).

amount of heat that can be transferred from the hot source to the cold source in a refrigerator. This parameter is related to the variation of magnetic entropy according to the following relationship [57]:

$$RCP = -\Delta S_{Max} \times \delta T_{FWHM}$$

Table 5 shows that the values of the maximum magnetic entropy and the cooling capacity (RCP) of the compounds in the solid solution $Nd_2Fe_{17-x}Co_x$ (x = 0 and x = 2) increase with the copper content.

Indeed, the values of the magnetic entropy of the $Nd_2Fe_{17-x}Co_x$ system are close to those determined in the R_2Fe_{17} system (R: Gd, Tb, Dy and Er) [58], as well as in solid solutions $Pr_2(Fe, Al)_{17}$ and $(Pr, Dy)\ _2Fe_{17}$ [59].

7. Magnetic and magnetocaloric properties of the solid solution $Gd_2Fe_{17-x}Cu_x$

7.1 Magnetic properties of the $Gd_2Fe_{17-x}Cu_x$ series ($0 \leq x \leq 1.5$)

Magnetic measurements of the temperature dependence of the magnetization were performed using a DSM-8 Differential Magneto /Susceptometer, in the temperature range from 300 K up to 600 K and under a weak applied magnetic field of order 0.12 T. From the M (T) curves of the series of compounds $Gd_2Fe_{17-x}Cu_x$ (x = 0; 0.5; 1 and 1.5) shown in **Figure 14**.

The study of the evolution of the magnetization as a function of the temperature and of the M (H, T) field carried out on either side of the Curie temperature with a step of 5 K for each compound of the solid solution $Gd_2Fe_{17-x}Cu_x$ (x = 0; 0.5; 1 and 1.5). These magnetization isotherms clearly show that our samples exhibit paramagnetic behavior for high temperatures and ferromagnetic behavior for temperatures below the T_C. At low temperatures ($T < T_C$), the curves show a rapid increase in magnetization for a field $H < 1$ T and as soon as the applied magnetic field increases, the magnetization tends to saturate. We note that with a field of the order of 1 T, the studied compounds reach saturation. For ($T > T_C$), the magnetization curves as a function of the magnetic field applied at different temperatures become more and more linear. **Figure 15** shows the isothermal magnetization curves measured under a magnetic field variation of 0–2 T, of the compounds of the solid solution $Gd_2Fe_{17-x}Cu_x$ (x = 0; 0.5; 1 and 1.5).

The Arrott method allows us to study the order of the magnetic transition of $Gd_2Fe_{17-x}Cu_x$ samples ($0 \leq x \leq 1.5$) using the isotherms M (H, T). For a second-order transition, the Arrott curves have positive slopes while for a first-order transition, the curves are negative. The Arrott isotherms presented in **Figure 16** for the $Gd_2Fe_{17-x}Cu_x$ samples (x = 0; 0.5; 1 and 1.5) show positive slopes which reveals the presence of a second order magnetic transition for all the samples.

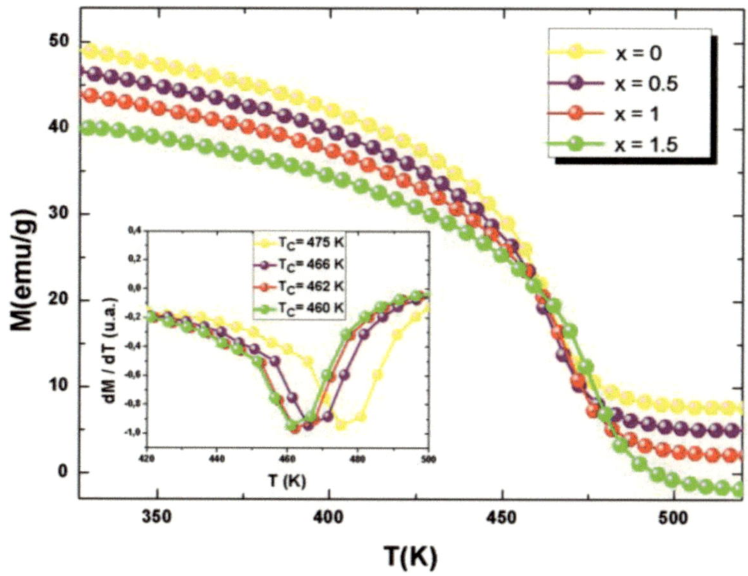

Figure 14.
Evolution of the magnetization as a function of the temperature M (T) for $Gd_2Fe_{17-x}Cu_x$ (x = 0; 0.5; 1 and 1.5) under a field of 0.12 T.

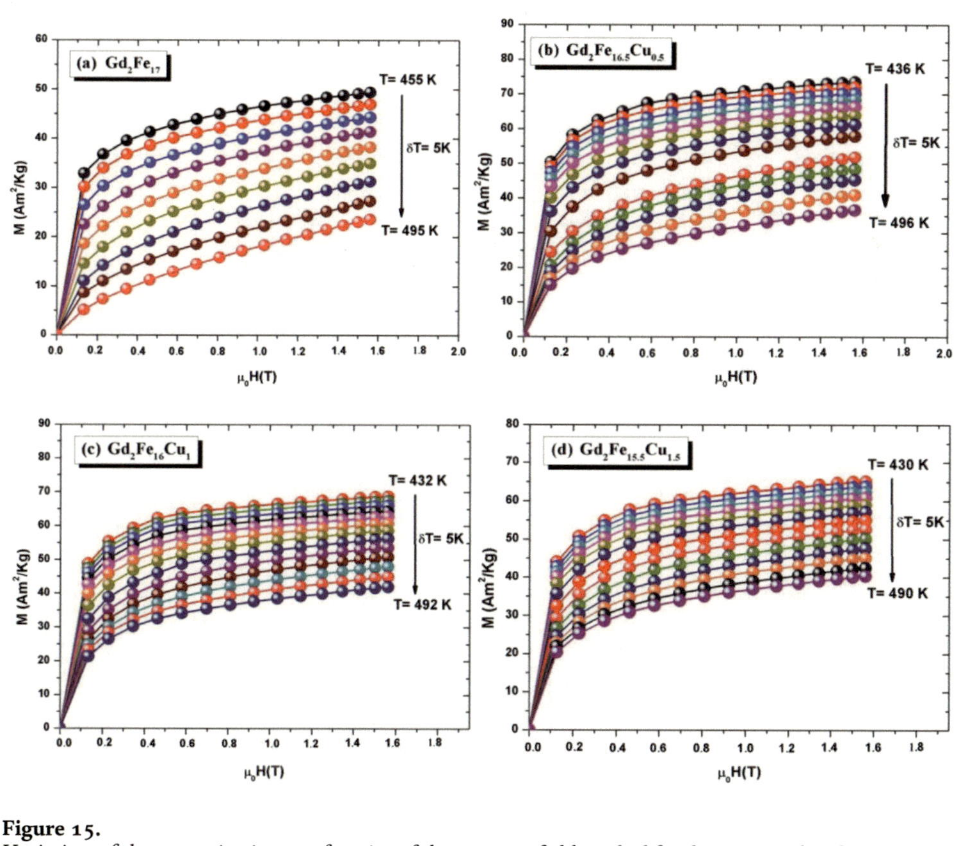

Figure 15.
Variations of the magnetization as a function of the magnetic field applied for the compounds $Gd_2Fe_{17-x}Cu_x$ (x = 0; 0,5; 1 et 1,5).

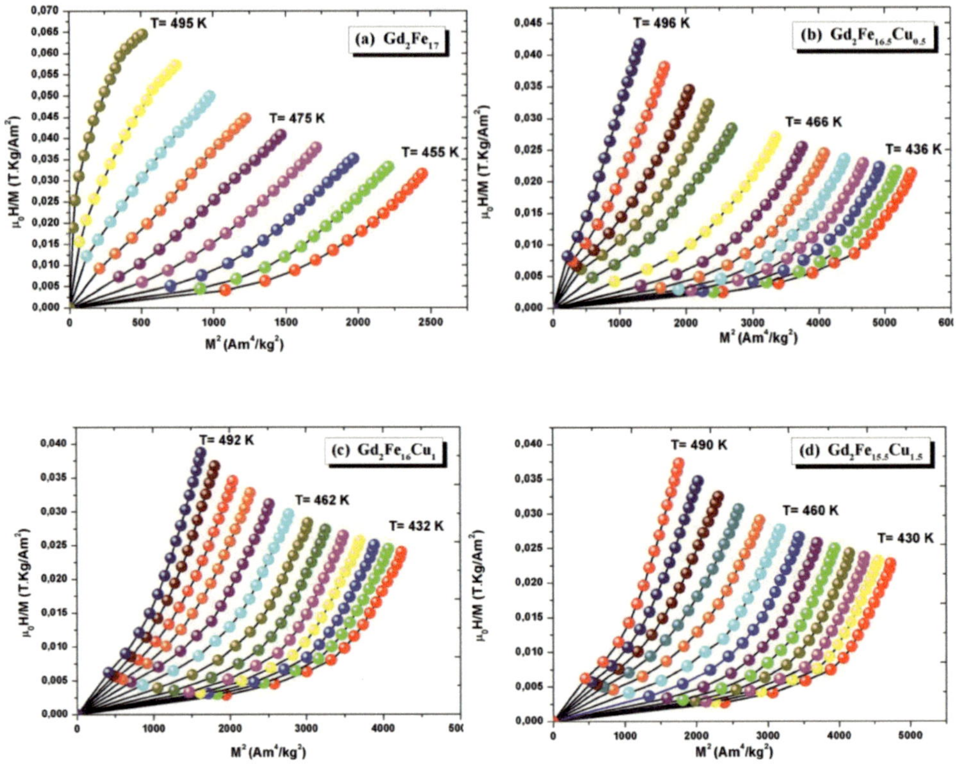

Figure 16.
Arrott plots of compounds $Gd_2Fe_{17-x}Cu_x$ (x = 0; 0,5; 1 et 1,5).

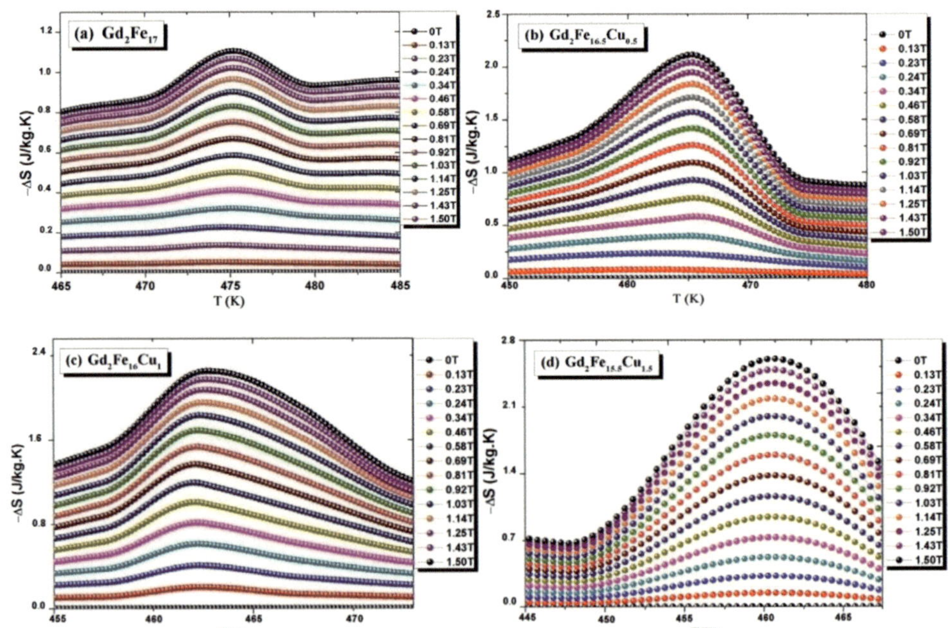

Figure 17.
Variations in magnetic entropy as a function of temperature and the applied magnetic field of the $Gd_2Fe_{17-x}Cu_x$ compounds (x = 0; 0.5; 1 and 1.5).

7.2 Magnetocaloric effect of the $Gd_2Fe_{17-x}Cu_x$ series ($0 \le x \le 1.5$)

The variation in magnetic entropy ($-\Delta S_{Max}$) as a function of temperature under different magnetic fields for $Gd_2Fe_{17-x}Cu_x$ compounds (x = 0; 0.5; 1 and 1.5) is shown in **Figure 17**. The curves show a maximum in the vicinity of the Curie temperature. It is found that the substitution of iron by copper causes an increase in the magnetocaloric effect. The values of the magnetic entropy of the $Gd_2Fe_{17-x}Cu_x$ system are close to those determined in solid solutions $Pr_2(Fe, Al)_{17}$ and $Gd_2(Fe, Si)_{17}$ [59].

Table 6 shows that the values of the maximum magnetic entropy and the cooling capacity (RCP) of the compounds in the solid solution $Gd_2Fe_{17-x}Cu_x$ ($0 \le x \le 1.5$) increase with the copper content.

X	0	0.5	1	1.5
ΔS_{Max} $(J \cdot kg^{-1} \cdot K^{-1})$	1,14	2,14	2,35	2,54
RCP $(J.\ kg^{-1})$	18	23,7	27,2	31,5

Table 6.
Magnetic entropy and cooling capacity of $Gd_2Fe_{17-x}Cu_x$ compounds (x = 0; 0.5; 1 and 1.5).

8. Conclusion

In summary, the structural characterization, determination of the magnetic properties and the magnetocaloric effects were performed for the $Nd_2Fe_{17-x}Co_x$ (x = 0; 1; 2; 3 and 4) and $Gd_2Fe_{17-x}Cu_x$ (x = 0, 0.5, 1 and 1.5) compounds. A single rhombohedral Th2Zn17 phase was obtained after one week of heat treatment at 800°C for synthesis by means of arc-melting. The crystal structure of the parent R_2Fe_{17} was found to remain unchanged for the $Gd_2Fe_{17-x}Cu_x$ and $Sm_2Fe_{17-x}Ni_x$ compounds which crystallize in the rhombohedral structure with the space group $R\bar{3}m$. Knowledge of the Maxwell relationship is essential to understanding the behavior of magnetocaloric materials and they also serve as indicators in the search for ever better performing compounds. Thus, a lot of information is to be extracted:

The magnetization decreases as the temperature increases, $(\partial M/\partial T)\ H < 0$, so ΔS_{Max} should be negative and this is well confirmed for our samples.

For ferromagnetic compounds, the maximum value of$|(\partial M/\partial T)\ H|$is reached at T_C, which means that the evolution of ΔS_{Max} as a function of temperature describes a peak whose maximum is at T_C, which works well with the results found for our systems. In all the materials studied in the thesis, the behavior of ΔS_{Max} gradually decreases on either side of the order temperature. The Curie temperature of $Nd_2Fe_{17-x}Co_x$ (x = 0; 1; 2; 3 and 4) compounds increases with Co content from 326 to 620 K for x = 0 to x = 4), respectively. The Curie temperature is the result of two effects: a magneto-volumic effect linked to Fe-Fe distances and an electronic effect linked to the filling of the 3d band of iron. For the system studied, the volume of the mesh remains constant as a function of the degree of substitution of the cobalt, this shows that the electronic effect dominates the magneto-volume effect. The study of the magnetic properties shows the Curie temperature and the entropy variation ΔS increase. This increase is explained by the Co–Co interactions which become stronger.

The experimental results show that in the $Gd_2Fe_{17-x}Cu_x$ (x = 0, 0.5, 1 and 1.5) compounds, the Curie temperature decreases by increasing the copper content from 475 K for (x = 0) to 460 K for (x = 1.5). This behavior can be attributed to the unit-cell volume decrease and the magnetic dilution. Moreover, we found that the copper substitution leads to an increase in ΔS_{Max} from 1.14 J/kg for Gd_2Fe_{17} to 2.54 J/kg K for $Gd_2Fe_{15.5}Cu_{1.5}$.

Author details

Mosbah Jemmali[1,2*] and Lotfi Bessais[3]

1 LSME, Faculty of Science, University of Sfax, Sfax, Tunisia

2 Department of Chemistry, College of Science and Arts, Ar-rass, Qassim University, Buraydah, Saudi Arabia

3 Univ Paris Est Creteil, CNRS, ICMPE, Thiais, France

*Address all correspondence to: jmosbah73@gmail.com

References

[1] V. K. Pecharsky, K. A. Gschneidner, Phys. Rev. Lett., 78 (1997) 4494.

[2] V. K. Pecharsky, K. A. Gschneidner, J. Mag. Mag. Mat., 167 (1997) 179.

[3] V. K. Pecharsky, K. A. Gschneidner, Phys. Lett., 70 (1997) 3299.

[4] D. M. Dimiduk, Y. W. Kim, R. Wagner, M. Yamaguchi. TMS, Warrendale, (1995) 3.

[5] N. S. Stolo, C. T. Liu, S. C. Deevi, Intermetallics., 8 (2000) 1313.

[6] K. H. J. Buschow, Handbook of Mag netic Materials., 10 (1997) 463.

[7] L. Bessais, C. Djega-Mariadassou, V. H. Ky, N. X. Phuc, J. Alloys Compd. 426 (2006) 22.

[8] E. Burzo, A. Chelkovski, and H.R. Kirchmayr, Landolt-Bornstein Handbook (Berlin, 1990).

[9] S. Chikazumi, Physics of Ferromagnetism, 2nd edn. (Oxford: Oxford University Press, 1997).

[10] P.J. Cregg and L. Bessais, J. Magn. Magn. Mater., 202 (1999) 554.

[11] D. Givord and R. Lemaire, IEEE Trans. Magn. , 10 (1974) 109.

[12] K.H.J. Buschow, Rep. Prog. Phys.,40 (1977) 1179.

[13] K.H.J. Bushow, Handbook of Magnetic Materials, vol. 4(Amsterdam: Elsevier, 1988).

[14] E. Belorizky, M.A. Fremy, J.P. Gavigan, D. Givord, and H.S.Li, J. Appl. Phys., 61 (1987) 3971.

[15] K.H.J. Buschow, Rep. Prog. Phys., 54 (1991) 1123.

[16] J.J.M. Franse, R.J. Radwanski, and K. H.J. Buschow,Handb. Magn. Mater., 7 (1993) 307.

[17] K.H.J. Buschow, Magnetism and Processing of Permanent Magnet, Volume 10 of Handbook of Magnetic Materials(Amsterdam: Elsevier, 1997).

[18] J.X. Zhang, L. Bessais, C. Djega-Mariadassou, E. Leroy, and A. Percheron-Guegan, Appl. Phys. Lett., 80 (2002)1960.

[19] L. Bessais, E. Dorolti, and C. Djega-Mariadassou, Appl. Phys. Lett., 87 (2005) 192503.

[20] H. Chen, Y. Zhang, J. Han, H. Du, Ch. Wang, and Y. Yang,J. Magn. Magn. Mater. , 320 (2008) 1382.

[21] P. Alvarez, P. Gorria, V. Franco, J.S. Marcos, M.J. Perez,J.L.S. Llamazares, I. P. Orench, and J.A. Blanco, J. Phys. Condens. Matter ,22 (2010) 216005.

[22] P. Alvarez, P. Gorria, J.S. Marcos, J. L.S. Llamazares, andJ.A. Blan, J. Phys. Condens. Matter 25, (2013) 496010.

[23] R. Guetari, R. Bez, A. Belhadj, K. Zehani, A. Bezergheanu,N. Mliki, L. Bessais, and C.B. Cizmas, J. Alloys Compd., 588 (2014) 64.

[24] S. Charfeddine, K. Zehani, L. Bessais, and A. Korchef, J.Solid State Chem., 238 (2016) 15.

[25] C. Djega-Mariadassou and L. Bessais, J. Magn. Magn.Mater. 210.

[26] M. Saidi, S. Walha, K. Nouri, A. Kabadou, M. Jemmali, L. Bessais, Journal of Alloys and Compounds 781 (2019) 159.

[27] N. Bouchaala, M. Jemmali, K. Nouri, S. Walha, A. BenSalah, L. Bessais, J. Phase Equilib. Diffus., 38 (2017) 561.

[28] N. Bouchaala; M. Jemmali; T. Bartoli; K. Nouri; I. Hentech; S. Walha; A. Ben Salah; L . Bessais. Journal of Solid State Chemistry, 258 (2018) 501

[29] M. Saidi, K. Nouri, S. Walha, E. Dhahri, A. Kabadou, M. Jemmali and L. Bessais, Journal of Electronic Materials, 48 (2019) 2242.

[30] K. Nouri, M. Jemmali, S. Walha, K. Zehani, L. Bessais, A. Ben Salah J. Alloys Compd. 661 (2016) 508.

[31] L. Bessais, C.D. Mariadassou, D.K. Tung, V.V. Hong, and N.X. Phuc.pp. J. Alloys Compd.,455 (2008) 35.

[32] I. Nehdi, M. Abdellaoui, C. D. Mariadassou, L. Bessais, H. Zarrouk. Physical and Chemical News,13 (2003) 21.

[33] C. Djéga.Mariadassou, L. Bessais, A. Nandra, J. M. Grenèche, E. Burzo. Phys. Rev. B,65 (2001) 14419.

[34] I.A. Al.Omari, S.S. Jaswal, A.S. Fernando, D.J. Sellmyer. J. Appl. Phys., 76 (1994) 6159.

[35] X.C. Kou, R. Grossinger, T.H. Jacobs, and K.H.J. Buschow. J. Magn. Magn. Mater., 88 (1990) 1.

[36] M. Jemmali, S. Walha, M. Pasturel, O. Tougait, R. Ben Hassen, H. Noël. J. Alloys Compd., 489 (2010) 421.

[37] M. Jemmali, S. Walha, , R. Ben Hassen, H. Noël. Asian Journal of Chemistry, 28 (2016), 1330.

[38] Z. W. Li., X.Z. Zhou and A.H. Morrish, Phys. Rev. B, 51 (1995) 2891.

[39] C. Lin, Y.X. Sun, Z.X. Liu, H.W. Jiang and Z. X. Liu, H.W. Jiang and Z. H. Liu, IEEE Trans. Magn., 28(1 992) 2844.

[40] PhD thesis by A. Nandra, defense 2003, structural and magnetic study of nanocrystalline alloys Sm_2Fe_{17-x} Si_x and their non-equilibrium $SmFe_{9-y}Si_y$ precursors, non-carburized and carburized.

[41] Z.Y. Ren, W.Y.Lee, C.D. Qin, D.H. L. Ng and X.Y.Ma, J. Appl.Phys. , 85 (1999) 4672.

[42] F.M. Yang, W Gong and G.C. Hadjipanayis, J. Appl.Phys. ,76 (1994) 6156.

[43] D.M. Zhang, Y. H. Gao, B.M. Yu, C. Q. Tang, N. Tang, X.P. Zhong, W.C. Lin, F.M.Yang and F.R. de Boer, J. Appl. Phys. , 76 (1994) 7452.

[44] E. Burzo, Solid State Commun, 89 (1994) 519.

[45] P. Brommer, Physica B,154 (1989) 197.

[46] H.Liu, D.Wang, S.Tang, Q. Cao, T. Tang, B. Gu, Y. Du, J. Alloys Compd. 346 (2002) 314.

[47] X. B. Liu, Z. Altounian, J. Magn. Magn. Mater., 292 (2005) 83.

[48] G. V. Brown, J. Appl. Phys., 47 (1976) 3673.

[49] V. K. Pecharsky, K. A. Gschneidner. Jr, Phys. Rev. Let., 78 (1997) 4494.

[50] D. Givord and R. Lemaire. IEEE Trans.Magn., MAG-10 (1974)109.

[51] C. Djega-Mariadassou and L. Bessais. J.Magn.Magn.Mater., 210 (2000)81.

[52] Z.W. Li and A. H.Morrish. Phys. Rev. B, 55(1997)3670.

[53] B. G. Shen, F.W.Wang, H. Y. Gong, Z. H. Cheng, B. Liang, and J. X. Zhang. J. Phys. :Condens. Matter, 7 (1995) 883.

[54] F. M. Yang, X.W. Li, N. Tang, J.L. Wang, Z. H. Lu, T. Y.Zhao, Q. A. Li, J. P.

Liu, and F. R. de Boer. J. Alloys Compd., 221 (1995)248.

[55] D. Givord and D. Courtois, J.Magn. Magn. Mater. , 196 (1999) 684.

[56] M.Z. Huang, W.Y.Ching and Z.Q. Gu J. Appl.Phys. , 81 (1997) 5112.

[57] A. M.Tishin, K. A. Gschneidner Jr. and V. K. Pecharsky. Phys. Rev. B, 59 (1999) 503.

[58] H. Chen, Y.Zhang, J. Han, H. Du, Ch. Wang, Y.Yang.Journal of Magnetism and Magnetic Materials 320 (2008) 1382.

[59] K. Zehani, R. Guetari, N. Mliki, L. Bessais. Physics Procedia, 75(2015) 1435